Marcel Baracchi-Frei

Real-time GNSS Software Receiver

Marcel Baracchi-Frei

Real-time GNSS Software Receiver
Optimized for General Purpose Microprocessors

Südwestdeutscher Verlag für Hochschulschriften

Impressum/Imprint (nur für Deutschland/only for Germany)
Bibliografische Information der Deutschen Nationalbibliothek: Die Deutsche Nationalbibliothek verzeichnet diese Publikation in der Deutschen Nationalbibliografie; detaillierte bibliografische Daten sind im Internet über http://dnb.d-nb.de abrufbar.
Alle in diesem Buch genannten Marken und Produktnamen unterliegen warenzeichen-, marken- oder patentrechtlichem Schutz bzw. sind Warenzeichen oder eingetragene Warenzeichen der jeweiligen Inhaber. Die Wiedergabe von Marken, Produktnamen, Gebrauchsnamen, Handelsnamen, Warenbezeichnungen u.s.w. in diesem Werk berechtigt auch ohne besondere Kennzeichnung nicht zu der Annahme, dass solche Namen im Sinne der Warenzeichen- und Markenschutzgesetzgebung als frei zu betrachten wären und daher von jedermann benutzt werden dürften.

Coverbild: www.ingimage.com

Verlag: Südwestdeutscher Verlag für Hochschulschriften GmbH & Co. KG
Dudweiler Landstr. 99, 66123 Saarbrücken, Deutschland
Telefon +49 681 37 20 271-1, Telefax +49 681 37 20 271-0
Email: info@svh-verlag.de

Approved by: Neuchâtel, University, Dissertation, 2010

Herstellung in Deutschland:
Schaltungsdienst Lange o.H.G., Berlin
Books on Demand GmbH, Norderstedt
Reha GmbH, Saarbrücken
Amazon Distribution GmbH, Leipzig
ISBN: 978-3-8381-2870-2

Imprint (only for USA, GB)
Bibliographic information published by the Deutsche Nationalbibliothek: The Deutsche Nationalbibliothek lists this publication in the Deutsche Nationalbibliografie; detailed bibliographic data are available in the Internet at http://dnb.d-nb.de.
Any brand names and product names mentioned in this book are subject to trademark, brand or patent protection and are trademarks or registered trademarks of their respective holders. The use of brand names, product names, common names, trade names, product descriptions etc. even without a particular marking in this works is in no way to be construed to mean that such names may be regarded as unrestricted in respect of trademark and brand protection legislation and could thus be used by anyone.

Cover image: www.ingimage.com

Publisher: Südwestdeutscher Verlag für Hochschulschriften GmbH & Co. KG
Dudweiler Landstr. 99, 66123 Saarbrücken, Germany
Phone +49 681 37 20 271-1, Fax +49 681 37 20 271-0
Email: info@svh-verlag.de

Printed in the U.S.A.
Printed in the U.K. by (see last page)
ISBN: 978-3-8381-2870-2

Copyright © 2011 by the author and Südwestdeutscher Verlag für Hochschulschriften GmbH & Co. KG and licensors
All rights reserved. Saarbrücken 2011

"Positioning is like oil drilling. Close is not good enough."
unknown author

Contents

Contents   iii

Acronyms   ix

Nomenclature   xiii

1 Introduction   1
- 1.1 Context and background 1
- 1.2 Satellite navigation 2
  - 1.2.1 History 2
  - 1.2.2 Overview of satellite navigation systems 3
  - 1.2.3 GPS 3
  - 1.2.4 GLONASS 4
  - 1.2.5 Galileo 4
  - 1.2.6 Compass 5
- 1.3 Thesis outline 5
- 1.4 Publications 7

2 Introduction to GPS receivers   9
- 2.1 Introduction 9
- 2.2 GPS signal characteristics 10
  - 2.2.1 C/A Code 11
  - 2.2.2 Doppler Frequency Shift 12
  - 2.2.3 Navigation Data 12
  - 2.2.4 Transmitted signal 13
  - 2.2.5 Received power level 14
- 2.3 Sampling frequency 15
- 2.4 Acquisition 17
  - 2.4.1 Serial search architecture 19
  - 2.4.2 Parallel frequency search architecture 21
  - 2.4.3 Parallel code search architecture 22
- 2.5 Code and carrier tracking 24
  - 2.5.1 Code tracking 25
  - 2.5.2 Carrier tracking 27
- 2.6 Calculating the position 29
  - 2.6.1 Positioning basics 29

		2.6.2	Basic equations for calculating the position	31
		2.6.3	Measurement of pseudorange	32
		2.6.4	User position from pseudoranges	33
		2.6.5	Position with more than four satellites	34
	2.7	Summary		36

3 Software receivers — 37

- 3.1 Introduction — 37
- 3.2 History — 37
- 3.3 Importance of software receivers — 38
- 3.4 Definition and types — 40
- 3.5 An ideal software receiver — 41
- 3.6 Challenges — 41
  - 3.6.1 Data rate — 42
  - 3.6.2 Signal sample conversion — 43
  - 3.6.3 Baseband processing — 44
- 3.7 Existing solutions and their complexity — 46
  - 3.7.1 Baseband processing architecture — 46
  - 3.7.2 Carrier generation and mixing algorithms — 48
  - 3.7.3 Code generation and mixing algorithms — 54
  - 3.7.4 Serial search architecture — 59
  - 3.7.5 Parallel code search architecture — 59
  - 3.7.6 Parallel frequency search architecture — 61
  - 3.7.7 Comparison of the different acquisition methods — 62
  - 3.7.8 FFT performance values — 63
- 3.8 Alternate processing methods — 66
  - 3.8.1 Single Instruction Multiple Data (SIMD) — 67
  - 3.8.2 Bitwise (vector processing) operations — 68
  - 3.8.3 Use of Graphics Processing Unit (GPU) — 69
- 3.9 Summary — 70

4 New architecture and algorithms for a software receiver — 71

- 4.1 Introduction — 71
- 4.2 Separation hardware / software — 71
- 4.3 New architecture — 72
  - 4.3.1 RF front-end — 72
  - 4.3.2 Baseband Pre-Processing — 73
  - 4.3.3 Host interface — 73
  - 4.3.4 Hardware prototype — 76
- 4.4 New algorithms — 77
  - 4.4.1 General concept of the baseband processing — 78
  - 4.4.2 Carrier generation and mixing — 79
  - 4.4.3 Code generation and mixing — 81
  - 4.4.4 Parallel frequency search — 84
  - 4.4.5 Parallel code search — 86
  - 4.4.6 Final software receiver architecture — 87

| | | 4.5 Summary | 91 |

5 Implementation of new architecture and algorithms — 93
- 5.1 Introduction — 93
- 5.2 Front-end unit — 93
  - 5.2.1 RF front-end board — 94
  - 5.2.2 FPGA — 94
  - 5.2.3 USB interface — 96
- 5.3 USB real-time data handling — 97
  - 5.3.1 Finding device — 98
  - 5.3.2 Initialization — 98
  - 5.3.3 Events — 99
  - 5.3.4 Threads — 99
  - 5.3.5 Notes — 101
- 5.4 Baseband processing — 102
  - 5.4.1 Computation of the partial sums — 102
  - 5.4.2 Code generation — 103
  - 5.4.3 Carrier generation — 104
  - 5.4.4 Code removal — 106
  - 5.4.5 Carrier removal — 107
  - 5.4.6 Acquisition — 108
  - 5.4.7 Re-Acquisition — 110
  - 5.4.8 Tracking — 111
  - 5.4.9 Code phase measurements — 113
  - 5.4.10 Tracking loops — 114
- 5.5 Aiding — 115
- 5.6 Final software receiver prototype — 116
- 5.7 External libraries and header files — 117
  - 5.7.1 FFTW — 117
  - 5.7.2 External header files — 118
- 5.8 Summary — 118

6 Setup, tests, and results — 121
- 6.1 Introduction — 121
- 6.2 Test setup — 122
- 6.3 Description of the figures — 124
- 6.4 Results — 126
  - 6.4.1 Accuracy and performance tests (simulated signals) — 126
  - 6.4.2 Accuracy and performance tests (real signals) — 132
  - 6.4.3 Requirement tests — 133
- 6.5 Summary — 136

7 Conclusion and Outlook — 141
- 7.1 Conclusion — 141
- 7.2 Outlook — 142

A Carrier-to-Noise Density — 145
- A.1 Power in decibels (dBm, dBW) — 145
- A.2 Signal-to-Noise Ratio (SNR) — 145
- A.3 Thermal Noise — 146
- A.4 Carrier-to-Noise Density (C/N0) — 147

B Phase-Locked Loops — 149
- B.1 Basic Phase-Locked Loop — 149
- B.2 First order Phase-Locked Loop — 152
- B.3 Second order Phase-Locked Loop — 153
- B.4 Transform from continuous to discrete systems — 155

C Signal description baseband architectures — 159
- C.1 Real baseband architecture — 160
- C.2 Complex baseband architecture — 162

D Platform specifications — 165
- D.1 Asus EeePC 1000H (Notebook) — 165
- D.2 Dell Latitude D430 (Notebook) — 166
- D.3 Dell Precision 380 (Single core desktop) — 166
- D.4 Dell Precision 380 (Dual core desktop) — 167
- D.5 Custom made PC (Single core desktop) — 167

Bibliography — 169

Dedicated to all persons working in the GNSS domain.
Hope this thesis gives you some inputs.

Acronyms

ADC	Analog-to-Digital Converter
BBPP	Baseband Pre-Processing
BOC	Binary Offset Carrier
BPSK	Binary Phase Shift-Key
C/A	Coarse Acquisition
CDMA	Code Division Multiple Access
DFT	Discrete Fourier Transform
DLL	Delay-Locked Loop
DSP	Digital Signal Processor
ESA	European Space Agency
ESPLAB	Electronic and Signal Processing LABoratory
FDMA	Frequency Division Multiple Access
FFT	Fast Fourier Transform
FIR	Finite Impulse Response
FLL	Frequency-Locked Loop
FPGA	Field Programmable Gate Array
GDOP	Geometric Dilution Of Precision
GIOVE	Galileo In-Orbit Validation Element
GLONASS	GLObal NAvigation Satellite System
GNSS	Global Navigation Satellite System
GPS	Global Positioning System
GPU	Graphics Processing Unit

HOW	Hand-Over Word
IC	Integrated Circuit
IF	Intermediate Frequency
IIR	Infinite Impulse Response
IMT	Institute of Microtechnology
LO	Local Oscillator
LORAN	LOng RAnge Navigation
LSBs	Least Significant Bits
MBOC	Multiplexed Binary Offset Carrier
MEO	Medium Earth Orbit
MMX	Multi Media Extension
MSBs	Most Significant Bits
NAVSAT	Navy Navigation Satellite System
NCO	Numerically Controlled Oscillator
NF	Noise Figure
PaC-SDR	Parameter-controlled SDR
PGA	Programmable Gain Amplifier
PLL	Phase-Locked Loop
PRN	Pseudo-Random Noise
PVT	Position, Velocity, and Time
RF	Radio Frequency
RHCP	Right Hand Circular Polarization
SAR	Search And Rescue
SDR	Software-Defined Radio
SIMD	Single Instruction Multiple Data
SISD	Single Instruction Single Data
SNR	Signal to Noise Ratio
SR	Software Receiver
TLM	Telemetry
TOA	Time Of Arrival

TOW	Time Of Week
TTFF	Time To First Fix
UHF	Ultra High Frequency
UTC	Universal Time Coordinated
VCO	Voltage Controlled Oscillator

Nomenclature

Operators and functions

$\otimes$	XOR operator
$\lceil \cdot \rceil$	Round operator
$\lfloor \cdot \rfloor$	Floor operator
$[\alpha]^{-1}$	Inverse of matrix α
$[\alpha]^T$	Transpose of matrix α
$sinc(x)$	$\sin(x)/x$

Constants

c	Speed of light = 299'792.458 km/s
i	Imaginary unit
k_B	Boltzmann constant $\approx 1.38 \cdot 10^{-23}$ J/K
π	Pi ≈ 3.14159

Architectures description

N	Number of points
N_{bin}	Number of Doppler bins
N_{ch}	Number of channels
N_{cor}	Number of correlators
N_ϕ	Number of code phases
N_s	Number of samples
N_q	Number of bits for signal quantization
n_s	Number of different code phases to be stored in LUT
F_s	Sampling frequency
$F_{s_{IQ}}$	Sampling frequency in complex architecture (= $F_s/2$)
F_{if}	Intermediate frequency
F_{code}	Frequency of the C/A code (1.023 MHz)

F_{car}	Generated carrier frequency by a NCO
F_{dop}	Doppler frequency
F_{qrz}	Quartz offset
F_{LO}	Frequency of the Local Oscillator
ΔF_{car}	NCO frequency resolution
Δf	Frequency step size or resolution
M	Number of sub-intervals of T_{int}
P	Number of consecutive partial correlations
T_{coh}	Pre-detection time
T_{int}	Integration time
$B(k)$	Carrier batch size for respective carrier phase
K	Number of carrier phases in one integration period
$P_L(j)$	Code partial sum size for chip j
$P(j)$	Code partial sum for chip j
$P_I(n)$	Code partial sum of number n for in-phase signal
$P_q(n)$	Code partial sum of number n for quadrature signal
U	Partial sum matrix
U'	Simplified partial sum matrix
V	Partial sum vector
Δn	Sample offset of code partial sum
$\overline{B}$	Average carrier batch size
$\overline{P}$	Average number of samples per code chip
$\sum[a:b]$	Partial sum from sample a to sample b
E_I, P_I, E_I	Output of in-phase correlators
E_Q, P_Q, L_Q	Output of quadrature correlators
I, Q	In-phase, quadrature signal
L_{dB}	Loss due to frequency mismatch
NCO_{inc}	NCO increment
T_{CPU}	CPU time to perform a computation
PR	Precision of code alignment
W	Bit-width of NCO
f_{OPS}	Number of FFT operations the CPU can perform
u	Sampling factor $= 2 \cdot BW/F_s$

Nomenclature

Signal description

A_P	Amplitude of P(Y) code
A_C	Amplitude of C/A code
A_{CA}	Amplitude of C/A code
BW_x	Bandwidth of the signal x
$C(t)$	C/A code
$D(t)$	Data code
$D(n)$	Data code, sampled
F_x	Frequency of the signal x
G	Processing gain
P_{noise}	Thermal noise floor
P_{signal}	Power of signal
P_b	Navigation data bit rate (50 Hz)
$P(t)$	P(Y) code
R_c	PRN chip rate (1.023 MHz)
S_{L1}	GPS L1 signal
T	Temperature [K]
f_0	Common GPS frequency (10.23 MHz)
$r_{ik}(m)$	Cross-correlation function between C_i and C_k
$r_{ss}(m)$	Autocorrelation function of $s(t)$
t	time
ω_{IF}	Angular frequency of Intermediate Frequency
ϕ	Initial phase or phase difference

FFT description

$X(k), Y(k)$	Discrete Fourier transform of $x(n), y(n)$
$X^*(k)$	Complex conjugate of $X(k)$
$z(n)$	Circular cross correlation
$Z(n)$	Discrete Fourier transform of $z(n)$
N	Size of the FFT
M	Number of different correlation values
$x(n), y(n)$	Finite length sequences

Loop description

B_n	Equivalent noise bandwidth
$F(s)$	Filter function in the s-domain
$F(z)$	Filter function in the z-domain

$H(s)$	Transfer function in the s-domain
$H(z)$	Transfer function in the z-domain
$H_e(s)$	Error transfer function
k_0	Gain of the amplifier in loop
k_1	Gain of the VCO
$u(t)$	Unit step function
V_0	Input voltage to the VCO
$\epsilon(s)$	Error function
$\theta_i(t)$	Input signal of the loop
$\theta_f(t)$	Output signal of the loop
$\theta_f(s)$	Laplace transform of $\theta_f(t)$
ω	Angular frequency
ω_0	Center angular frequency of the VCO
$\omega_2(t)$	Output frequency of the VCO
ω_n	Natural frequency
ζ	Damping factor

Pseudorange and position computation

b_{ut}	User clock error
b_u	User clock bias error
S_a	Position of satellite a
t_{si}	Time of emission of the signal
t_u	Time of reception of the signal
t'_{si}	Actual satellite clock time
t'_u	Actual user clock time
x_a	Distance to satellite a
Δb_i	Satellite clock error
ρ	Pseudorange

Chapter 1

Introduction

This chapter puts the performed work in a global context and gives some information about the organization of the document.

The first section of this chapter explains the background and the context for developing and implementing a real-time software receiver on a general purpose microprocessor (Section 1.1). Then, a short introduction to the different Global Navigation Satellite System (GNSS) is given (Section 1.2), followed by an outline of this thesis with the description of the different chapters (Section 1.3). The last section gives an overview of the publications made during the research activities (Section 1.4).

Even though the final solution focuses on GPS, the other GNSS are nevertheless presented for completeness as most of the presented algorithms could be adopted to them.

1.1 Context and background

Research activities at the Electronic and Signal Processing LABoratory (ESPLAB) of the Institute of Microtechnology (IMT) in the domain of GNSS started back in 1997 with a project in collaboration with Asulab (http://www.asulab.com), the research and development laboratory of Swatch Group (http://www.swatchgroup.com). The goal of the project was to design an Integrated Circuit (IC) of a Global Positioning System (GPS) receiver that could be embedded into a watch. The objective of the project was achieved, but unfortunately, the watch was never commercialized.

Since then, several research projects were carried out in the field of GNSS, among others a dual-frequency GPS L1/L2 receiver and a Galileo receiver for Search And Rescue (SAR) applications. This allowed our research team to develop a deep and funded know-how of the different fields in GNSS development, among others the design of specific and optimized Radio Frequency (RF) front-ends and the optimized implementation of the baseband processing on Field Programmable Gate Array (FPGA) and embedded systems.

In 2007, the ESPLAB was contacted by u-blox AG (http://www.u-blox.com) for developing and implementing a real-time capable software receiver on a general purpose microprocessor. The goal of the project was to investigate the feasibility of running a complete GPS L1 (and eventually Galileo) receiver in software with the same performance as a commercial (hardware) receiver. If this study was successful, the new solution was foreseen to be implemented in software. Even though it looked not so complicated in the beginning, some challenges were quickly identified. One of them consisted in the fact that – by the time – all the developed receivers at ESPLAB did the most computational demanding operations on a FPGA. If the well known and proven architectures were just "copied" to software, a real-time implementation would not be possible (see Section 3.6 for more details). But nevertheless, some other institutes and research teams had already shown – at that time – running solutions (even though sometimes quite limited) of software receivers proving that there were possibilities and a huge research potential.

1.2 Satellite navigation

Satellite navigation is a method using a GNSS to accurately identify the position and the time anywhere on Earth. With the help of a GNSS, the following values can be determined with the following accuracy [1]:

1. Exact position (longitude, latitude, and altitude) with an accuracy of between 20 m and approximately 1 mm.
2. Exact time (Universal Time Coordinated (UTC)) with an accuracy of between 60 ns to approximately $3 \cdot 10^{-13}$ s.

This section will give a very short overview of the different satellite navigation systems. It begins with a short history of how everything started and what were the predecessors of the current systems. Then, a short overview of the systems that are currently available and planned for the near future is given. The different systems are very briefly presented without going into technical details.

1.2.1 History

Where on Earth am I? The need for knowing his own position has been present ever since the human being started to explore the world.

Early predecessors of satellite navigation systems are the ground based DECCA, LOng RAnge Navigation (LORAN), and Omega systems which use terrestrial longwave radio transmitters instead of satellites. These positioning systems broadcast a radio pulse from a known "master" location, followed by repeated pulsed from a number of "slave" stations. The delay between the reception and sending of the signal at the slaves was carefully controlled, allowing the receivers to compare the delay between reception and the delay between sending. From this the distance to each of the slaves could be determined, providing a position fix.

The first satellite based navigation system was Transit (also known as Navy Navigation Satellite System (NAVSAT)), a system deployed by the U.S. military in the 1960s. Transit's operation was based on the Doppler effect: the satellites traveled on well-known paths and broadcasted their signals on a well known frequency. The received frequency slightly differs from the broadcast frequency because of the movement of the satellite with respect to the receiver. By monitoring this frequency shift over a short time interval, the receiver could determine its location to one side or the other of the satellite, and several such measurements combined with a precise knowledge of the satellite's orbit could determine a particular position.

1.2.2 Overview of satellite navigation systems

The domain of satellite navigation is evolving today at a high pace. At the time of writing this thesis, the United States NAVSTAR GPS is announced to be modernized with new modulations and services, the Russian GLObal NAvigation Satellite System (GLONASS) is in the process of being restored to full operation, the European Union's Galileo is foreseen to be operational in early 2014, and the People's Republic of China has indicated it will expand its regional Beidou navigation system into the global Compass navigation system by 2015.

Table 1.1 shows a comparison of the different GNSS systems.

	GPS	**GLONASS**	**Galileo**	**Compass**
Country	United States	Russia	Europe	China
Channel access	CDMA	FDMA	CDMA	CDMA
Orbital height	20'200km	19'100km	23'222km	21'150km
Period	12.0h	11.3h	14.1h	12.6h
Satellites	> 24	> 20	> 27	> 30
Status	operational	operational, CDMA in preparation	in preparation	in preparation

TABLE 1.1: Overview of GNSS systems

1.2.3 GPS

GPS was created and realized by the U.S. Department of Defense (DoD) in 1973. It consists of up to 32 Medium Earth Orbit (MEO) satellites in six different orbital planes, with the exact number of satellites varying as older satellites are retired and replaced. Operational since 1987 and globally available since 1994, GPS is currently the world's most utilized satellite navigation system (civilian and military use). All satellites have highly synchronized on-board Rubidium or Cesium atomic clocks as a frequency reference and broadcast Code Division Multiple Access (CDMA) ranging codes and navigation data currently on three frequencies, L1 (1575.42 MHz), L2 (1227.6 MHz), and L5 (1176.45 MHz), depending on the satellite generation. Every satellite has its own ranging code with low cross-correlation properties, but all satellites transmit on the same carrier frequencies (L1 and L2 or L1, L2 and L5).

As the system is already operational for almost 25 years, some modernizations are planned. One of the main objective of the next GPS modernization is the improvement of the quality of the civilian service by providing additional signals with new modulation types. It is planned to a modernized L1C/A signal which will be called L1C (see [2] for more details).

The signal characteristics are given in Section 2.2.

For details about the GPS system, please refer to the numerous textbooks, e.g., [3] and [4].

1.2.4 GLONASS

The GLONASS program was first started by the former Soviet Union and is today under the jurisdiction of the Commonwealth of Independent States (CIS). The first three test-satellites were launched into orbit on October 12, 1982. GLONASS has been a fully functional navigation constellation but since the collapse of the Soviet Union it has fallen into disrepair, leading to gaps in coverage and only partial availability. The Russian Federation has pledged to restore it to fully global availability with the help of India, who is participating in the restoration project. By April 2010, it is practically restored (21 out of 24 satellites are operational).

The biggest difference compared to the GPS (or also to the Galileo) system is that the channel access method is based on Frequency Division Multiple Access (FDMA), i.e., every satellite sends exactly the same code but on a different frequency. On the next generation of GLONASS satellites (GLONASS-K), also a CDMA signal is foreseen to be broadcasted [5].

More details can be found in [1] and directly on the homepage of the Federal Space Agency Information-Analytical Centre [6].

1.2.5 Galileo

The European Union (EU) and the European Space Agency (ESA) agreed on March 2002 to introduce their own alternative to GPS, called the Galileo positioning system, today scheduled to be working in 2014 [7]. The first validation satellite was launched on 28 December 2005 (GIOVE-A) and currently, two Galileo In-Orbit Validation Element (GIOVE) are available. The program took several delays, mainly because of financial questions.

The Galileo system is expected to improve the precision of the user position (for the open service a precision of approximately 4 to 15 m is expected) [1]. This is achieved through the application of a new modulation type, called Binary Offset Carrier (BOC) and Multiplexed Binary Offset Carrier (MBOC).

For details about the Galileo system, please refer to the numerous textbook, e.g., [3] and [8].

1.2.6 Compass

China has indicated they intend to expand their regional navigation system, called Beidou or Big Dipper, into a global navigation system; a program that has been called Compass in China's official news agency Xinhua. The Compass system is proposed to utilize 30 MEO satellites and five geostationary satellites. Having announced they are willing to cooperate with other countries in Compass' creation, it is unclear how this proposed program impacts China's commitment to the Galileo system (China joined in 2003 the Galileo project).

1.3 Thesis outline

This section will give a short outline of the document. The main objective of this thesis was to develop and implement a real-time capable software receiver on a general purpose microprocessor. For this mean, a completely new GNSS receiver architecture has been developed, implemented and tested.

Chapter 2 presents a short and general introduction to GPS receivers. This includes the description of the classical architecture of a GNSS receiver, the GPS signal characteristics and the different baseband processing blocks. A short section about calculating the user position is given at the end.

Chapter 3 introduces the term *software receiver* by first giving a short historical overview of the most important development milestones and by describing the importance and the impact of the new concept, together with a definition of the term *software receiver* that will be used in this thesis. Some of the main challenges when developing a software receiver are given afterwards and the last section covers the existing solutions (algorithms and processing methods), also with respect to the needed computational power.

Chapter 4 presents the new architecture of the software receiver that was developed during the work. This includes a discussion about where to split the receiver architecture between hardware and software and the resulting consequences. The new architecture and the algorithms are afterwards described in greater detail and ultimately, the final software receiver architecture is presented.

Chapter 5 presents the implementation of the architecture described in Chapter 4. The different elements and blocks are described in more detail, together with the whole environment that is necessary to run a software receiver on a standard microprocessor.

Chapter 6 presents the tests and the results of the implementation described in Chapter 5. This includes the test setup and the description of the different tests, but also the limitations of the current implementation. At the end of that chapter, the results for the accuracy and the requirement tests are presented.

Finally, Chapter 7 concludes the thesis.

1.4 Publications

This section gives an overview of the papers and journal articles published during the elaboration of this thesis.

Patent

- Clemens Bürgi, Marcel Baracchi, Grégoire Waelchli. A method of processing a digital signal derived from a direct-sequence spread spectrum signal. European Patent, Application No. 09405207.3 – 2411, February 2009.
- Clemens Bürgi, Marcel Baracchi, Grégoire Waelchli. A method of processing a digital signal derived from a direct-sequence spread spectrum signal and a receiver. U.S. Patent, Application No. 12/694,145, January 2010.

Journal

- Marcel Baracchi-Frei, Grégoire Waelchli, Cyril Botteron, and Pierre-André Farine. Real-Time GNSS Software Receivers: Challenges, Status, and Perspectives. Article in *GPS World*, pp. 40–47, September 2009.
- Grégoire Waelchli, Marcel Baracchi-Frei, Cyril Botteron, and Pierre-André Farine. Distributed Arithmetic for Efficient Baseband Processing in Real-time GNSS Software Receivers. Article in *Research Letters in Signal Processing, Hindawi Publishing Corporation*, November 2009.
- Marcel Baracchi-Frei, Grégoire Waelchli, Cyril Botteron, and Pierre-André Farine. Real-Time GNSS Software Receivers: Challenges, Status, and Perspectives. Article in *MyCoordinates*, pp. 7–9, May 2010.

Conference

- Cyril Botteron, Grégoire Waelchli, Giuseppe Zamuner, Marcel Frei, Davide Manetti, Frédéric Chastellain, Pierre-André Farine, and Patrice Brault. A Flexible Galileo E1 Receiver Platform for the Validation of Low Power and Rapid Acquisition Schemes. In *ION GNSS*, Forth Worth, TX, USA, September 2009.
- Frédéric Chastellain, Cyril Botteron, Grégoire Waelchli, Marcel Frei, Davide Manetti, Pierre-André Farine, and Patrice Brault. A Galileo E1b,c RF front-end for Search-and-Rescue Applications. In *ENC GNSS*, Geneva, Switzerland, 2007.
- Grégoire Waelchli, Giuseppe Zamuner, Marcel Frei, Frédéric Chastellain, Elham Firouzi, Cyril Botteron, Pierre-André Farine, and Patrice Brault. Development, implementation, and validation of a real-time Galileo E1-B signal acquisition and tracking scheme. In *ENC GNSS*, Geneva, Switzerland, 2007.
- Marcel Frei, Davide Manetti, Elham Firouzi, Grégoire Waelchli, Cyril Botteron, Pierre-André Farine, Nicolas Voirol, Patrick Weber, and Jean-Pierre Aubry. Study of a Novel OCXO Characterization Based on Oven Supply Currents for Enhanced Holdover Compensation. In *ETFT, Toulouse Space Show '08*, Toulouse, France, 2008.
- Marcel Baracchi-Frei, Grégoire Waelchli, Cyril Botteron, and Pierre-André Farine. Real-Time GNSS Software Receivers: Challenges, Status, and Perspectives. In *ENC GNSS*, Naples, Italy, 2009.
- Grégoire Waelchli, Marcel Baracchi-Frei, Cyril Botteron, and Pierre-André Farine. Real-time Carrier Generation for a GNSS Software Receiver. In *International Symposium on GPS/GNSS*, ICC Jeju, Korea, 2009.
- Grégoire Waelchli, Marcel Baracchi-Frei, Cyril Botteron, and Pierre-André Farine. Performances of a New Correlation Algorithm for a Platform-Independent GPS Software Receiver. In *ION ITM*, Anaheim, CA, USA, 2009.
- Marcel Baracchi-Frei and Grégoire Waelchli. Real-Time GNSS Software Receiver: Challenges, Status, and Perspectives. In *Navigare '09*, Neuchatel, Switzerland, 2009.

Chapter 2

Introduction to GPS receivers

2.1 Introduction

This chapter describes the classical architecture of a GNSS receiver, starting with the signal characteristics of the GPS system and then going through the different baseband processing units. The chapter will be closed with a short description of the methods for calculating the position. Most of the information is extracted from [3] and [9]. Further details can be found in the above mentioned references and in [4], [8], [10], and [11].

2.2 GPS signal characteristics

In this section, the properties of the GPS satellite signals are described, including frequency assignments, modulation format, and a short overview of the content of the navigation data.

The GPS signals are transmitted currently on three radio frequencies on the Ultra High Frequency (UHF) band [12]. These frequencies are referred to as L1, L2, and L5 and are derived from a common frequency, $f_0 = 10.23$ MHz, as given in Equation 2.1.

$$\begin{aligned} f_{L1} &= 154 \cdot f_0 = 1575.42 \text{ MHz} \\ f_{L2} &= 120 \cdot f_0 = 1227.60 \text{ MHz} \\ f_{L5} &= 115 \cdot f_0 = 1176.45 \text{ MHz} \end{aligned} \quad (2.1)$$

Two other frequencies have been proposed to extend and improve the system. This consists of the frequency L3 (1381.05 MHz) used by the Nuclear Detonation (NUDET) Detection System Payload (NDS) to detect signals of nuclear detonations and other high-energy infrared events and L4 (1379.913 MHz) for additional ionospheric corrections. In this document, only the L1 frequency will be considered.

The GPS L1 signal that will be transmitted by the satellites is composed of the three components given in Table 2.1.

Component	Description
Carrier	The carrier wave with the frequencies f_{L1} as described in Equation 2.1.
Spreading sequence	Each satellite has two unique spreading sequences or codes (C/A and P(Y)) as described below.
Navigation data	The navigation data contains information regarding satellite orbits and transmission time and has a bit rate of 50 bps. This information is uploaded to all satellites from the ground stations in the GPS control segment.

TABLE 2.1: Components forming the GPS signal

The Coarse Acquisition (C/A) code is a sequence of 1023 chips (a chip corresponds to a bit but is called *chip* to emphasize that it does not hold any information). The code is repeated each millisecond giving a chipping rate of 1.023 MHz.

The encrypted P(Y) code is a longer code (more than 2^{42} chips) with a chipping rate of 10.23 MHz. It repeats itself each week starting at Saturday/Sunday midnight (which is the beginning of the GPS week).

The C/A code is only modulated onto the L1 carrier while the P(Y) code is modulated onto both the L1 and L2 carrier. The aforementioned modulation method is illustrated in Figure 2.1.

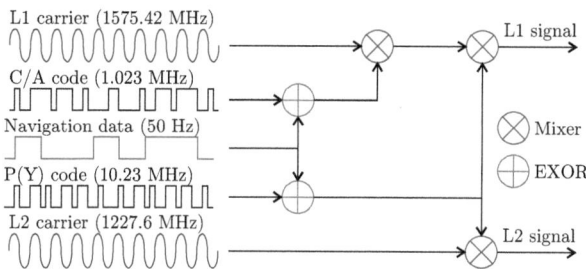

FIGURE 2.1: GPS signal modulation

The final signal shows a Binary Phase Shift-Key (BPSK) modulation where the carrier is instantaneously phase-shifted by 180° at the time of a chip change. When a navigation bit transition occurs, the phase of the resulting signal is also phase-shifted by 180°.

2.2.1 C/A Code

The spreading codes used as C/A codes in GPS belong to a unique family of sequences and are often referred to as *Gold codes*, as described by Robert Gold in 1967. They are also referred to as Pseudo-Random Noise (PRN) sequences or codes, because of their deterministic characteristics with noise-like properties. More details and the technical description how to generate them can be found in [10] and [11].

The most important characteristic – the correlation properties – will shortly be discussed here:

- *Nearly no cross correlation*: All the C/A codes are nearly uncorrelated with each other, i.e., the cross correlation for two codes C^i and C^k of satellite i and k can be written as:

$$r_{ik}(m) = \sum_{l=0}^{1022} C^i(l) \cdot C^k(l+m) \approx 0 \qquad \text{for all } m$$

- *Nearly no correlation except for zero lag*: All C/A are nearly uncorrelated with themselves, except for zero lag. This property makes it easy to find out when two similar codes are perfectly aligned. The autocorrelation property for the satellite k can be written as

$$r_{kk}(m) = \sum_{l=0}^{1022} C^k(l) C^k(l+m) \approx 0 \qquad \text{for } |m| \geq 1$$

2.2.2 Doppler Frequency Shift

As the signals are transmitted by moving satellites and received by (possibly also moving) receivers, they are affected by a Doppler frequency shift. This effect has an impact on both the acquisition and the tracking of the GNSS signals. For a stationary GNSS receiver, the maximum Doppler frequency shift for the L1 frequency is around ± 5 kHz (due to the satellite motion). The receiver motion also creates a Doppler frequency shift of approximately 1.46 Hz per each 1 km/h.

The Doppler frequency shift affects both the carrier frequency and the C/A code. The effect on the C/A code is small because of the low chip rate of the C/A code which is 1572.42 MHz/1.023 MHz (i.e., 1540 times lower than the L1 carrier frequency). It follows that the Doppler frequency shift on the C/A code is 3.2 Hz and 6.4 Hz for the stationary and the high-speed GNSS receiver, respectively. This Doppler frequency shift on the C/A code can lead to a misalignment between the received and the locally generated PRN code that decreases the correlation result.

2.2.3 Navigation Data

The navigation data message is transmitted with the bit rate of 50 bps. Figure 2.2 shows the overall structure of the entire navigation message. The basic format of the navigation data is a 1500 bit long frame containing 5 subframes, each having a length of 300 bits. Every subframe contains 10 words of a length of 30 bits. Subframes 1, 2, and 3 are repeated in each frame while the last subframes, 4 and 5, have 25 different versions (always with the same structure, but different content) referred to as page 1 to 25. With the bit rate of 50 bps, the transmission of a subframe lasts 6 seconds, one frame lasts 30 seconds, and one entire navigation message lasts 12.5 minutes.

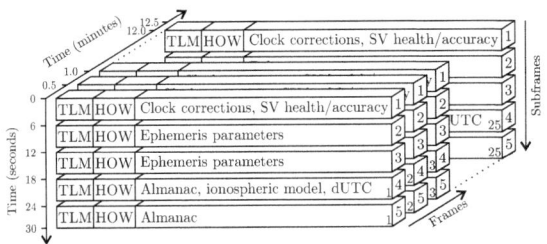

FIGURE 2.2: GPS navigation data structure
(Figure: Frank van Diggelen)

The subframes always begin with two special words, the Telemetry (TLM) and the Hand-Over Word (HOW). *TLM* is the first word of each subframe and is repeated every 6 seconds. It contains a 8 bit preamble followed by 16 reserved bits and parity. The preamble is used for frame synchronization.

Chapter 2 Introduction to GPS receivers 13

HOW contains a 17 bit truncated version of the Time Of Week (TOW), followed by two flags supplying information to the user of antispoofing, etc. The next three bits indicate the subframe ID that shows in which of the five subframes this HOW is located.

In addition to the TLM and HOW, each subframe contains eight words of data. The following table gives a short description of the different subframes. More details (and the exact description of the bits) can be found in [12].

Subframe 1 - Satellite Clock and Health Data
  The first subframe contains mainly clock information. That is information needed to compute at what time the navigation message is transmitted from the satellite. Additionally, subframe 1 contains health data indicating whether or not the data should be trusted.

Subframes 2 and 3 - Satellite Ephemeris Data
  Subframes 2 and 3 contain the satellite ephemeris data. This information relates to the satellite orbit and is needed to compute the exact satellite position.

Subframes 4 and 5 - Support Data
  Subframes 4 and 5 contain almanac data of all satellites (every satellite broadcasts the almanacs for all GPS satellites, but only ephemeris data for itself). Almanacs are the ephemerides and clock data with reduced precision. The remainder of subframes 4 and 5 contain various data, e.g., UTC parameters, health indicators, and ionospheric parameters.

2.2.4 Transmitted signal

The final L1 signal transmitted from the satellite k is given in Equation 2.2.

$$\begin{aligned} S_{L1} = &A_P \cdot P(t) \cdot D(t) \cdot cos(2 \cdot \pi \cdot f_{L1} \cdot t + \phi) \\ &+ A_C \cdot C(t) \cdot D(t) \cdot sin(2 \cdot \pi \cdot f_{L1} \cdot t + \phi) \end{aligned} \quad (2.2)$$

where S_{L1} is the signal at the frequency L1, A_P is the amplitude of the P(Y) code, $P(t) = \pm 1$ represents the P(Y) code, $D(t) = \pm 1$ represents the data code, f_{L1} is the L1 frequency, ϕ is the initial phase, A_C is the amplitude of the C/A code, and $C(t) = \pm 1$ represents the C/A code.

Since the L1 carrier is essentially modulated with a 1.023 MHz PRN sequence, the frequency domain representation of the L1 signal looks like a *sinc* centered at the GPS L1 frequency. The main lobe, having a bandwidth of 2.046 MHz, contains more than 90% of the signal energy. At first glance, it might seen wasteful that GPS occupies at least 2 MHz of spectrum in order to transmit the navigation data at a rate of 50 Hz. But the 1.023 MHz PRN sequence allows for very desireable signal characteristics:

- Transmission on the same frequency;
- Precise ranging;
- Processing gain due to de-spreading of PRN code;

- Rejection of reflected signals;
- Anti-jamming properties.

2.2.5 Received power level

After modulation, the L1 signal is transmitted with a Right Hand Circular Polarization (RHCP). The received power level is approximately -130 dBm on the Earth's surface with a 3 dBi antenna. With a 2 MHz bandwidth and at room temperature, the thermal noise floor P_{noise} is given in Equation 2.3.

$$\begin{aligned} P_{noise} &= k_B \cdot T \cdot \text{BW}_{GPS_{L1}} \\ &= (1.38 \cdot 10^{-23}\,\frac{\text{J}}{\text{K}}) \cdot (294\text{ K}) \cdot (2 \cdot 10^6\text{ Hz}) \\ &= -111\text{ dBm} \end{aligned} \quad (2.3)$$

where k_B is the Boltzmann constant, T the antenna temperature in Kelvin, and $\text{BW}_{GPS_{L1}}$ the spectral bandwidth of the GPS L1 C/A signal in Hz. This means that the received GPS signal is approximately 19 dB below the thermal noise floor and the signal can only be recovered by de-spreading the signal (i.e., removing the PRN code).

The correlation with the correct PRN code results in a processing gain defined as the ratio of the PRN chip rate R_c to the data bit rate R_b, as given in Equation 2.4. For the GPS L1 C/A signal, the PRN rate is 1.023 MHz and the data rate is 50 Hz, resulting in a processing gain of 43 dB.

$$G = 10 \cdot \log_{10}\left(\frac{R_c}{R_b}\right) = 10 \cdot \log_{10}\left(\frac{1.023\text{ MHz}}{50\text{ Hz}}\right) = 43\text{ dB} \quad (2.4)$$

With a processing gain of 43 dB, a typical received GPS signal has a pre-detection Signal to Noise Ratio (SNR) of 20-30 dB, as given in Equation 2.5.

$$\begin{aligned} \text{SNR}_{dB} &= 10 \cdot \log_{10}\left(\frac{P_{signal}}{P_{noise}}\right) = P_{signal,dB} - P_{noise,dB} \\ &= (-130\text{ dBm} + 43\text{ dB}) - (-111\text{ dBm}) \\ &= 24\text{ dB} \end{aligned} \quad (2.5)$$

In Equation 2.5, a perfect RF front-end with a Noise Figure (NF) of 0 dB was used for the calculation. If a more realistic case of a receiver with a NF of 3 dB is considered, the SNR_{dB} decreases to 21 dB.

Appendix A gives an overview of the different RF quantities and units commonly used in discussions of GPS signal power levels.

2.3 Sampling frequency

The choice of the sampling frequency is directly related to the C/A code chip rate and is a quite important aspect related to the front-end selection. This section will discuss the choice of the sampling frequency and the resulting impact on the distance resolution.

The C/A code chip rate is 1.023 MHz and the sampling frequency should not be a multiple number of this chip rate, i.e., the sampling frequency should not be synchronized with the C/A code rate. For example, working at a sampling frequency of 4.092 MHz (4 · 1.023 MHz) is not a good choice for the following reasons:

○ With this sampling rate, the time between two adjacent samples becomes 244 ns (1/4.092 MHz) and this time is used to determine the beginning of the C/A code.
○ The distance resolution becomes 73.31 m ($244 \cdot 10^{-9} \cdot 3 \cdot 10^8$ m) that is too coarse to obtain the desired accuracy of the user position.

Figure 2.3 shows the C/A code chip rate and the digitizing points, Figure 2.3(a) with synchronized sampling and Figure 2.3(b) with unsynchronized sampling. In each figure, there are two sets of digitizing points where the lower row is the time-shifted version of the upper row.

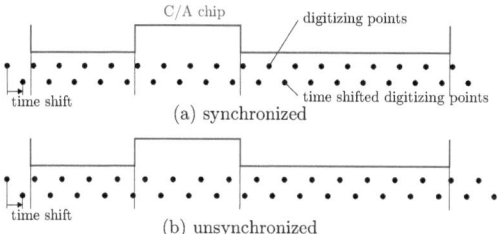

FIGURE 2.3: Synchronized and unsynchronized C/A code sampling

In Figure 2.3(a), the time shift is slightly less than 244 ns and it can be seen that the two sets of digitized data are exactly the same (i.e., same number of samples per C/A code chip). This illustrates that shifting by less than 244 ns produces the same output data if the sampling frequency is synchronized with the C/A code. Therefore, a finer time resolution than 244 ns cannot be derived through signal processing.

In Figure 2.3(b), the sampling frequency is slightly lower than 4.092 MHz and therefore no longer synchronized with the C/A code. The output data from the time shifted version is different from the original data as shown in the figure and therefore, a finer time resolution can be obtained by signal processing. This allows to determine the beginning of the C/A code more precisely.

Strictly speaking, the navigational performance (in the sense of a low variance of the estimated values) of a receiver is basically a function of the navigation autocorrelation function (and its derivatives) and the received signal power (C/N_0). The sample rate does not directly enter. Several options for choosing the sampling rates are possible: oversampling, Nyquist sampling, and sub-Nyquist sampling.

For the following short discussion, we assume a real-valued signal of dual-sided bandwidth BW. The signal is first bandpass filtered (also with a bandwidth of BW) and then (bandpass) sampled with a sampling frequency of F_s. The noise is received with an uniform power spectral density of N_0. The scheme is shown in Figure 2.4 [13].

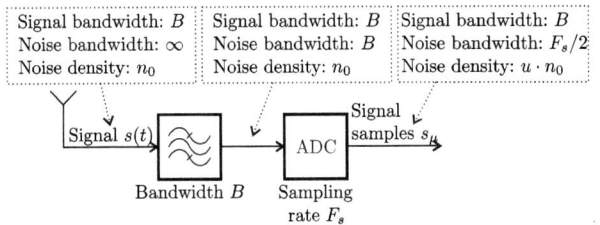

FIGURE 2.4: Noise bandwidth and density in RF front-end

From the definition of a correlation function (it is basically a time average over many signal samples), the important fact is that the expected value of a correlator is independent of the sampling rate. So, the sampling frequency does not affect the shape of the correlation function and the sampling also leaves the total noise power invariant. Before sampling, the total noise power is spread over a bandwidth BW, whereas after the ADC the total noise power is spread over $F_s/2$. In the case where $u = 2 \cdot BW/F_s$ is an integer, the noise-power spectral density after the ADC is flat [14].

Overall, the following sampling cases can occur:

○ $F_s > 2 \cdot BW$ – *Oversampling*: The complete band is sampled and, due to the oversampling, the noise is colored after the ADC. Optional spectral whitening produces white noise, but does not affect the signal. After spectral whitening, we can assume $u = 1$.

○ $F_s = 2 \cdot BW$ – *Nyquist sampling*: Signal and noise are sampled with the Nyquist rate. This is the ideal case and no losses occur. $u = 1$.

○ $F_s < 2 \cdot BW$ – *Sub-Nyquist sampling*: Aliasing of the noise occurs, but the signal autocorrelation function is not affected by the sub-Nyquist sampling. $u = 2 \cdot BW/F_s$.

When oversampling is used, the signal processing includes spectral regions which are not covered by the navigation signal. These regions are ignored when correlating the received signal samples with the navigation signal replicas. However, oversampling can be quite useful because the high working sample rate allows small shift of the replica signals (e.g., early or late signals) with respect to a reference signal (e.g., prompt signal). Typically, it is much easier to shift a signal than to do a full regeneration of the signals. In fact, in hardware receivers, high oversampling rates are used to achieve small correlator spacing.

Nyquist sampling represents a kind of ideal sampling; the signal is optimally captured, the noise after the ADC is white, and no noise-aliasing losses occur. A high-performance software receiver should work with the Nyquist sampling rate. Usually, finite analog-filter fall-off steepness effects can be ignored. The sampling rate is (twice) the 3-dB bandwidth.

Sub-Nyquist sampling is the method of choice if the computational load must be kept as low as possible. Shifting the reference signal is not needed but the signal power must be sufficiently high to cope with the noise-aliasing losses. Because of the noise aliasing, the effective C/N_0 value is reduced by the undersampling factor u: $C/N_0 \rightarrow \frac{1}{u} \cdot C/N_0$. A detailed discussion about sub-Nyquist sampling can be found in [14].

An important restriction for choosing the sample rate occurs if real-valued signals are used. In this case, the replica signal correlates with negative frequency components of the received signal. Those correlation values are required to average to zero during the correlation process. This implies that after the ADC the aliased center frequency should be larger than 0.

2.4 Acquisition [11]

This section describes the purpose of the acquisition stage and presents the different architectures that can be found in standard GNSS receivers. The main focus lies on their advantages and drawbacks and the description of the involved operations while the estimation of the required computational power is presented in Chapter 3.

The purpose of the acquisition stage is first of all to identify the visible satellites. Once the signal from a given satellite is found, two important parameters must be measured: the beginning of the C/A code period (code phase) and the carrier frequency of the input signal. These two parameters are passed afterwards to the tracking stage.

A set of collected data usually contains signals of several satellites. Each signal has a different C/A code with a different starting time and a different Doppler frequency. The satellites are distinguished by the PRN sequences as describes in Chapter 2.2.1. These PRN codes show a low cross-correlation and a high auto-correlation for zero lag and therefore allow to find the corresponding satellite in the received data stream.

The code phase is the time alignment of the PRN code of a specific satellite in the current data block. To be able to remove the spreading code from the incoming signal, it is necessary to exactly determine the start of the C/A code.

The carrier frequency corresponds normally to the Intermediate Frequency (IF) in the case of a down-conversion architecture, but as the velocity of the satellite causes a Doppler shift (see Section 2.2.2), the final carrier frequency is slightly shifted (up to ± 10 kHz). To generate the local carrier and to remove the residual Doppler frequency from the incoming signal, it is therefore important to know the exact frequency.

The length of the data used to perform the acquisition is always a subject of discussion. The use of a longer data record results in a higher SNR ratio and therefore in higher sensitivity. But using a long data record increases the time of calculation and complicates the design. The Doppler frequency shift on the C/A code and the question whether there is a navigation data bit transition present can limit the length of the used data record.

A navigation data bit transition will spread the spectrum and the output will no longer be a continuous wave signal. This spectrum spread will degrade significantly the acquisition result. Since the navigation data is 20 ms or 20 C/A codes long (for GPS L1), the maximum data record for acquisition should be 10 ms. If a data record longer than 10 ms is used for acquisition, special care has to be taken because in a set of 20 ms of data, only one data bit transition can occur. If the first 10 ms of data contain a data bit transition, the next 10 ms will not have one for sure. In actual acquisition, even if there is a phase transition caused by a navigation data in the input data, the spectrum spreading is not very wide. For example, if 10 ms of data are used for acquisition and there is a phase transition at 5 ms, the width of the peak spectrum is about 400 Hz ($2/(5 \cdot 10^{-3})$). As this peak usually can be detected, the beginning of the C/A code can be found.

The second limit of data length comes from the Doppler effect on the C/A code. If a perfect correlation peak is normalized to 1, the correlation peak decreases to 0.5 when a C/A code is off by half a chip (this corresponds to 6 dB decrease in amplitude). As the chip frequency is 1.023 MHz and the maximum Doppler shift expected on the C/A code is 6.4 Hz (as discussed in Chapter 2.2.2), it takes about 78 ms ($1/(2 \cdot 6.4)$) for the two frequencies different by 6.4 Hz to change by half a chip (this corresponds to the maximum possible C/A code misalignment). As this data length limit is much longer than the 10 ms found in the navigation data bit transition discussion, the main limitation comes from there.

The carrier frequency separation is also a factor that needs to be considered carefully. The Doppler frequency range to be searched is normally ±5 kHz (for fast moving receivers). It is important to determine correctly the frequency steps needed to cover this 10 kHz range as this value influences directly the frequency resolution and the complexity of the search algorithm. The frequency step is related to the data length used in the acquisition stage [15]. As seen before, there is no correlation at the output when the input signal and the locally generated complex signal are off by one (or more) cycle. A misalignment of less than one chip results in a partial correlation. It is arbitrarily chosen that the maximum frequency separation allowed between the two signals is 0.5 cycle. If the data record has a length of 1 ms, a 1 kHz signal will change by one cycle in this time. In order to keep the maximum frequency separation at 0.5 cycle in 1 ms, the frequency step should be 1 kHz resulting in the furthest frequency separation of 500 Hz between the input signal and the correlating signal. The frequency separation is also the inverse of the data length, which is the same as a conventional FFT result.

The number of needed operations in the acquisition process is not linearly proportional to the total number of data points. When the data length is increased from 1 ms to 10 ms, the number of operations required to perform the acquisition is increased more than 10 times. Therefore, the length of the data sequence (or the integration time) should be kept at a minimum if the speed of the acquisition is important. More details about the acquisition stage can be found in [3], [10], and [11].

The following sections describe three standard methods of acquisition architectures that can currently be found in GNSS receivers.

2.4.1 Serial search architecture

The serial search acquisition is a method that is often used in CDMA systems, such as GPS or Galileo. Figure 2.5 shows a block diagram for one channel of this method. A channel is the basic element for processing the incoming signal and is allocated to one specific satellite.

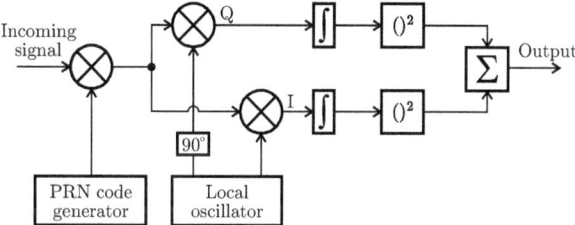

FIGURE 2.5: Serial search acquisition architecture

The PRN code generator produces a PRN sequence with a certain code phase (from 0 to 1022 chips) corresponding to the satellite that is acquired by this channel. The incoming signal is first multiplied with this sequence to remove the PRN code and afterwards with a locally generated carrier signal (with a specific frequency) to remove the Doppler offset. As shown in Figure 2.5, the multiplication with the local carrier frequency generates the in-phase signal I and the multiplication with a 90° phase-shifted version generates the quadrature signal Q. The I and Q signals are then integrated over the integration time T_{int} and finally squared and accumulated. Ideally, the complete signal power should be located on the I part of the signal, as the C/A code is only modulated onto that. However, if the phase of the received signal is not perfectly aligned with the phase of the locally generated carrier frequency, a part of the signal power can also be present on the Q part of the signal. It is necessary to investigate both the I and Q signal to be sure to receive the complete signal power and to detect the satellite.

The output of the acquisition stage is a value of the correlation between the incoming signal and the locally generated signals. If a predefined threshold is exceeded, the code phase and the frequency parameters are correct and can be passed to the tracking algorithms.

The serial search algorithm performs a two-dimensional sweep: a code phase sweep over all 1023 different code phases and a frequency sweep over all possible carrier frequencies of ± 5 kHz. The step size for the frequency search depends on the integration time, as given in Equation 2.6. For 1 ms integration time, the step size is roughly 500 Hz.

$$\Delta f = \frac{2}{3 \cdot T_{int}} \tag{2.6}$$

All in all, for an integration time of 1 ms this sums up to a total of

$$\underbrace{1023}_{\text{code phases}} \cdot \underbrace{\left(2 \cdot \frac{5'000}{500} + 1\right)}_{\text{frequencies}} = 1023 \cdot 21 = 41'943 \text{ combinations.} \tag{2.7}$$

The main weakness of this architecture is this very large number of combinations that have to be processed.

The PRN code can be generated either in real-time (with the use of a LFSR) or offline and then saved in memory for further use. While the first method reduces the amount of needed memory, it increases the complexity of the whole algorithm. A hardware receiver uses normally the real-time generation of the PRN codes, while for a software receiver the offline generation shows more advantages and the total amount of operations can be reduced. However, the serial search algorithm involves multiplication with all possible shifted versions of the PRN code and therefore, a total of 32'736 (= 32 satellites · 1023 code phases) different PRN codes must be saved in memory. The total amount of needed memory is even increased if the PRN codes have to be sampled at the sampling frequency of the front-end F_s (i.e., the PRN codes are saved in a so called *oversampled representation*). For a sampling frequency of 4 MHz, the length of the PRN sequence increases to 4'000 samples.

The local oscillator has to generate two carriers with a phase difference of 90°, corresponding to a cosine and a sine wave. The frequency of the generated carrier has to correspond to the IF ± the frequency step according to the examined frequency area. Again, the generation can be implemented in real-time (by the use of a Numerically Controlled Oscillator (NCO)) or the different frequencies can be calculated offline and saved in memory. For a software receiver, the offline generation offers more advantages and decreases the total amount of needed operations (see Section 3.7). The generated carrier frequency has to be sampled with the sampling frequency F_s of the front-end and has the length of the integration time T_{int}. As the available memory is normally limited, not all possible frequencies can be saved in memory. This leads to a possible mismatch between the locally generated carrier

Chapter 2 Introduction to GPS receivers

frequency and the received signal.

The last part of the serial search algorithm involves a squaring and integrating of the two results of the multiplications with the cosine and sine signals, respectively. The integration is simply a summation of all points corresponding to the length of the processed data (which corresponds to the integration time T_{int}). The squaring is introduced to obtain the total signal power and is performed on the results of the summation. The final step is to add the two values from the I and Q branches to obtain the total signal power. If the locally generated code is well aligned with the code of the incoming signal and the frequency of the locally generated carrier matches the frequency of the incoming signal, the output reaches its maximum.

2.4.2 Parallel frequency search architecture

The parallel frequency search architecture performs the search in the frequency domain [16]. This is done by bringing the incoming signal into the frequency domain by the mean of the Fourier transform [17]. Figure 2.6 shows a block diagram of the parallel frequency space search algorithm.

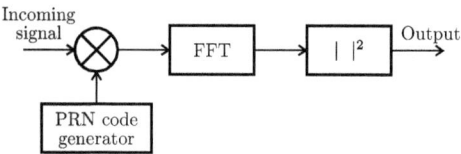

FIGURE 2.6: Parallel frequency search acquisition architecture

The incoming signal is first multiplied with the locally generated PRN sequence, with the code corresponding to a specific satellite and with a code phase between 0 and 1022 chips. The resulting signal is transformed into the frequency domain by a Fourier transform that can be either implemented as a Discrete Fourier Transform (DFT) or a Fast Fourier Transform (FFT). The FFT is the faster of the two, but it requires an input sequence with a radix-2 length, that is, 2^n, where n takes positive integer values.

If the incoming signal is perfectly aligned with the locally generated PRN sequence, the resulting signal after the multiplication is a continuous wave signal. This signal is afterward transformed into the frequency domain and the output of the Fourier transform will show a distinct peak in magnitude at the frequency of the carrier.

The accuracy of the determined frequency depends on the length of the DFT, i.e., on the number of samples in the analyzed data sequence. If 1 ms of data is analyzed, the number of samples can be found as 1/1000 of the sampling frequency F_s. With a DFT length of 10'000, the first $N/2$ output samples represent the frequencies from 0 to $F_s/2$ Hz. So, the resulting frequency resolution is given in Equation 2.8.

$$\Delta f = \frac{F_s/2}{N/2} = \frac{F_s}{N} \qquad (2.8)$$

Equation 2.9 gives the resolution for a sampling frequency of $F_s = 4$ MHz.

$$\Delta f = \frac{4 \text{ MHz}}{10'000} = 400 \text{ Hz} \qquad (2.9)$$

Where the serial search algorithm steps through all possible code phases and carrier frequencies, the parallel frequency search acquisition algorithm steps only through the 1023 different code phases, but for each code phase, a frequency domain transformation has to be calculated. The final speed of this architecture depends directly on the implementation of the frequency domain transformation.

The implementation of this method is quite straightforward as the algorithm can be implemented directly based on the block diagram shown in Figure 2.6. There are several FFT libraries that are optimized for different CPUs or DSPs. One example can be found in Section 5.7.1.

The first part of this method is identical to the first part of the serial search algorithm. The locally generated PRN code is multiplied with the incoming signal which is transformed afterwards into the frequency domain through the Fourier transform (DFT or FFT). If the locally generated code is perfectly aligned with the PRN code of the incoming signal, the output from the FFT will have a peak at the IF plus the Doppler offset frequency. If there is a misalignment between the two codes or if the locally generated PRN code is simply not present, the output of the FFT has a noise like shape and no peak can be seen.

2.4.3 Parallel code search architecture

This architecture performs a parallel code phase search over the 1023 possible values and reduces the number of different steps to 21 compared to 1023 in the parallel frequency search architecture. The calculation is performed with a circular cross correlation between the input and the PRN code without a shifted code phase. More details about this method can be found in [11].

The discrete Fourier transform of finite length sequences $x(n)$ and $y(n)$ both with the length N are computed as given in Equation 2.10.

$$\begin{aligned} X(k) &= \sum_{n=0}^{N-1} x(n) \cdot e^{-j \cdot 2 \cdot \pi \cdot k \cdot n / N} \\ Y(k) &= \sum_{n=0}^{N-1} y(n) \cdot e^{-j \cdot 2 \cdot \pi \cdot k \cdot n / N} \end{aligned} \qquad (2.10)$$

The circular cross-correlation sequence between two finite length sequences $x(n)$ and $y(n)$ both with length N and periodic repetition is computed as given in Equation 2.11.

$$\begin{aligned} z(n) &= \frac{1}{N} \cdot \sum_{m=0}^{N-1} x(m) \cdot y(m+n) \\ &= \frac{1}{N} \cdot \sum_{m=0}^{N-1} x(-m) \cdot y(m-n) \end{aligned} \quad (2.11)$$

The scaling factor $1/N$ will be omitted in the following equations to improve the readability. The discrete N-point Fourier transform of $z(n)$ can be expressed as given in Equation 2.12.

$$\begin{aligned} Z(k) &= \sum_{n=0}^{N-1}\sum_{m=0}^{N-1} x(-m) \cdot y(m-n) \cdot e^{-j \cdot 2 \cdot \pi \cdot k \cdot n / N} \\ &= \sum_{m=0}^{N-1} x(m) \cdot e^{-j \cdot 2 \cdot \pi \cdot k \cdot m / N} \cdot \sum_{n=0}^{N-1} y(m+n) \cdot e^{-j \cdot 2 \cdot \pi \cdot k \cdot (m+n)/N} \\ &= X^*(k) \cdot Y(k) \end{aligned} \quad (2.12)$$

where $X^*(k)$ is the complex conjugate of $X(k)$.

Figure 2.7 shows a block diagram of the parallel code phase search architecture. The incoming signal is multiplied with a locally generated carrier signal (giving the I signal) while the multiplication with the 90° phase-shifted version of the signal generates the Q signal. The I and Q signals are combined to form a complex input signal $x(n) = I(n) + j \cdot Q(n)$ to the DFT function. The PRN code is generated locally, transformed into the frequency domain and the result is complex conjugated.

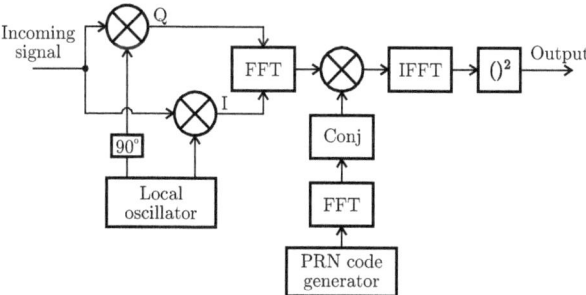

FIGURE 2.7: Parallel code search acquisition architecture

The Fourier transform of the input is multiplied with the Fourier transform of the PRN code. The result of the multiplication is transformed into the time-domain by an inverse Fourier transform. The absolute value of the output of the inverse Fourier transform represents the correlation between the input and the PRN code. If a peak is present in the correlation, the index of this peak marks the PRN code phase of the incoming signal.

Compared the the previous acquisition methods, the parallel code phase search acquisition method has cut down the search space to 41 carrier frequency bins. Additionally, the Fourier transform of the generated PRN code must only be calculated once for each acquisition (assuming that the Doppler frequency offset on the code can be neglected). As or each of the frequency bins, one Fourier transform and one inverse Fourier transform is performed, the computational efficiency of this method depends on the implementation of these functions.

In the same way as the other acquisition methods, the implementation is straightforward, as it can be implemented directly based on the block diagram shown in Figure 2.7.

2.5 Code and carrier tracking [8] [11]

In this chapter, the code and carrier tracking blocks of a GNSS receiver will be discussed, including the different discriminators with their advantages and drawbacks.

The theory of the Phase-Locked Loop (PLL) can be found in Appendix B.

The acquisition stage provides only rough estimates of the frequency and the code phase parameters of the satellite. The main purpose of the tracking stage is to refine these values, keep track, and demodulate the navigation data from a specific satellite.

The architecture of the tracking stage looks very similar to the acquisition stage, i.e., the incoming signal is multiplied with a locally generated carrier frequency and a locally generated PRN code. To be able to produce the two exact local signal replicas, some kind of feedback is needed that compares the generated replica with the incoming signal. The element that calculates this feedback is called *discriminator*.

A conventional PLL receives a continuous wave or a frequency-modulated signal as an input and the frequency of the VCO is controlled to follow the frequency of the input signal. In the case of a GNSS receiver, the input is the GNSS signal and the PLL must follow (or track) this signal. The GNSS signal is normally a bi-phase coded signal as the carrier and the code frequencies change due to the Doppler effect. Therefore, the C/A code information and the Doppler carrier frequency must be removed to track the GNSS signal. As a result, two PLL are required to track a GNSS signal. One is to track the C/A code and one is to track the carrier frequency. These two loops must be coupled together as shown in Figure 2.8 as an example.

Chapter 2 Introduction to GPS receivers 25

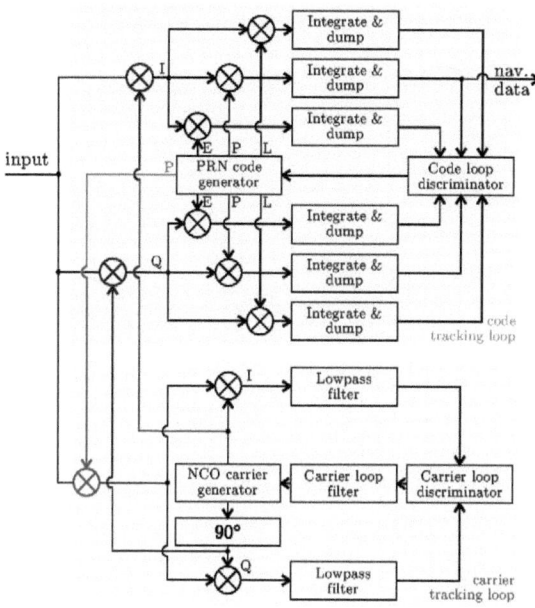

FIGURE 2.8: Code and carrier tracking loops

The code generator produces three outputs: an early, a late, and a prompt code. The prompt code (the one that should match best the incoming signal) is applied to the digitized input signal and strips the C/A code from the input signal (red mixer in Figure 2.8). Stripping off the C/A code means to multiply the locally generated C/A code (with the proper phase) with the input signal. The output will be a continuous wave signal with phase transitions caused only by the navigation data. This signal is applied to the input of the carrier tracking loop. The output from the carrier loop is a continuous wave with the carrier frequency of the input signal. This signal is in return used to strip the carrier from the digitized input signal, which means using this locally generated signal to multiply the input signal (blue mixer in Figure 2.8). The output is a signal with only the C/A code and no carrier frequency, which is applied to the input of the code tracking loop.

2.5.1 Code tracking

A code tracking loop is used to follow (or keep track of) a PRN code of a specific satellite. The output of such a code tracking loop is a perfectly aligned replica of the PRN sequence. In a GNSS receiver, the code tracking loop is usually a Delay-Locked Loop (DLL) called an early-late tracking loop. The idea behind a DLL is to correlate the input signal with three replicas of the code as shown in

Figure 2.8. The residual frequency is first removed by multiplying the incoming signal with a perfectly aligned local replica of the carrier wave (blue mixer in Figure 2.8). Afterwards, the signal is multiplied with three code replicas (early, prompt, and late) that have normally a spacing of ±1/2 chip. Then, the three outputs are integrated and dumped and the resulting output value indicates how much the specific code replica correlates with the code of the incoming signal.

The three correlation outputs (called E_I, P_I, and L_I) are compared to see which one provides the highest correlation value. Figure 2.9 shows an example of the code tracking.

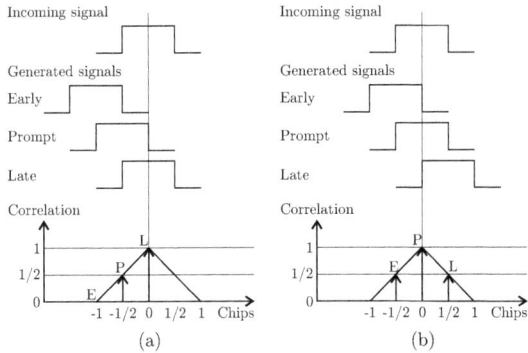

FIGURE 2.9: Code tracking (with 3 correlators)

In Figure 2.9(a) the late code has the highest correlation output value, so the code phase must be decreased. In Figure 2.9(b) the highest peak is found at the prompt replica and the early and late replicas show the same value, so the code phase is properly aligned. The configuration of a DLL with three correlators is optimal when the local carrier wave is perfectly locked in phase and frequency. When there is a small phase error on the local carrier wave, the signal will be more noisy and the energy of the input signal is distributed in the I and Q arms. Therefore, a design with six correlators (as shown in Figure 2.8) is often used that has the advantage to be less dependent of the phase of the local carrier wave. Table 2.2 gives an overview and a short description of some common DLL discriminators used in GNSS receivers.

The spacing between the correlators determines the noise bandwidth in the DLL. If the discriminator spacing is larger than 1/2 chip, the DLL would be able to handle wider dynamics and be more noise robust; on the other hand, a DLL with a smaller spacing would be more precise. It is also possible to implement an adjustable spacing width of the correlators so that if the signal-to-noise ratio suddenly decreases, the receiver uses a wider spacing in the correlators to be able to handle a more noisy signal. The advantage of this method is that a possible code lock loss can be avoided.

Type	Discriminator	Characteristics
Coherent	$E_I - L_I$	Simplest of all discriminators. Does not require the Q branch, but requires a good carrier tracking loop for optimal functionality.
Noncoherent	$(E_I^2 + E_Q^2) - (L_I^2 + L_Q^2)$	Early minus late power. The discriminator response is nearly the same as the coherent discriminator inside $\pm 1/2$ chip.
Noncoherent	$\dfrac{(E_I^2 + E_Q^2) - (L_I^2 + L_Q^2)}{(E_I^2 + E_Q^2) + (L_I^2 + L_Q^2)}$	Normalized early minus late power. The discriminator has a great property when the chip error is larger than $\frac{1}{2}$ chip; this will help the DLL to keep track in noisy signals.
Noncoherent	$P_I \cdot (E_I - L_I) + P_Q \cdot (E_Q - L_Q)$	Dot product. This is the only DLL discriminator that uses all 6 correlators.

TABLE 2.2: Types and characteristics of code tracking discriminators

2.5.2 Carrier tracking

A carrier tracking loop is used to follow (or keep track of) a frequency (that can be affected by a Doppler shift) of a specific satellite. The output of such a carrier tracking loop is a perfectly aligned carrier wave with the correct frequency and phase. Often a PLL or a Frequency-Locked Loop (FLL) are used. The problem with using an ordinary PLL is that it is sensitive to 180° phase shifts that occurs with a navigation bit transition. The solution is a so called *Costas loop*, as shown in Figure 2.8 (block carrier tracking loop) that is insensitive for phase transition. It uses two multiplications to generate an in-phase and a quadrature signal and it tries to keep all energy in the I (in-phase) branch. If it is assumed that the code replica in Figure 2.8 is perfectly aligned, the multiplication in the I branch yields to the Equation 2.13.

$$D(n) \cdot \cos(\omega_{IF} \cdot n) \cdot \cos(\omega_{IF} \cdot n + \phi) = \\ \frac{1}{2} \cdot D(n) \cdot [\cos(\phi) + \cos(2 \cdot \omega_{IF} \cdot n + \phi)] \quad (2.13)$$

where ϕ is the phase difference between the input signal and the local replica of the carrier.

The multiplication in the quadrature arms gives the Equation 2.14.

$$D(n) \cdot \cos(\omega_{IF} \cdot n) \cdot \sin(\omega_{IF} \cdot n + \phi) = \\ \frac{1}{2} \cdot D(n) \cdot [\sin(\phi) + \sin(2 \cdot \omega_{IF} \cdot n + \phi)] \quad (2.14)$$

If the two signals are low-pass filtered after the multiplication, the two terms with the double intermediate frequency ω_{IF} are eliminated and the two signals given in Equation 2.15 remain.

$$I = \frac{1}{2} \cdot D(n) \cdot \cos\phi,$$
$$Q = \frac{1}{2} \cdot D(n) \cdot \sin\phi. \qquad (2.15)$$

The phase error of the local carrier replica can be found as given in Equation 2.16.

$$\frac{Q}{I} = \frac{\frac{1}{2} \cdot D(n) \cdot \sin\phi}{\frac{1}{2} \cdot D(n) \cdot \cos\phi} = \tan(\phi),$$
$$\phi = \tan^{-1}\left(\frac{Q}{I}\right). \qquad (2.16)$$

The phase error is minimized when the correlation in the quadrature arm is zero and the correlation in the in-phase arm is maximum. The arctan discriminator is the most precise of the Costas discriminators, but also the most time-consuming. Table 2.3 lists other possible Costas discriminators.

Discriminator	Characteristics
$\text{sign}(I) \cdot Q$	The output is proportional to $\sin(\phi)$.
$I \cdot Q$	The output is proportional to $\sin(2 \cdot \phi)$.
$\tan^{-1}\left(\frac{Q}{I}\right)$	The output is the phase error.

TABLE 2.3: Types and characteristics of carrier tracking discriminators

Figure 2.10 shows a comparison of the responses of the three different discriminators. The phase discriminator outputs are computed using the expressions in Table 2.3 for all possible true phase errors. It can be seen that the discriminator outputs are zero when the real phase error is 0° and ±180°. This is why the Costas loop is insensitive to phase shifts of 180° in case of navigation data bit transition.

The behavior of a Costas loop when a phase shift of 180° occurs is illustrated more clearly in Figure 2.11 where the vector sum of I and Q is shown as the vector in the coordinate system. If the local carrier waves are in phase with the input signal, the vector would be perfectly aligned to the I-axis. But in the presence of a small error (φ), the vector also has a component on the Q-axis (as shown in the figure). When the tracking loops are working correctly, the vector sum of I and Q tends to remain aligned on the I-axis. This property ensures that if a navigation bit transition occurs, the vector on the diagram will flip by 180° (as indicated with the dashed blue vector in the figure).

The output of the phase discriminator is filtered to predict and estimate any relative motion of the satellite and to estimate the resulting Doppler frequency. More details can be found in [18], [3], [19].

Chapter 2 Introduction to GPS receivers

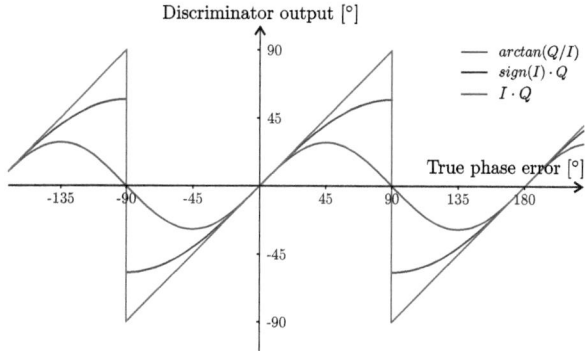

FIGURE 2.10: Comparison between common carrier tracking discriminators

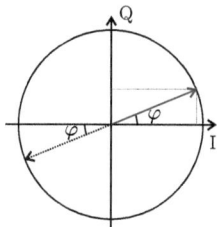

FIGURE 2.11: Effect of phase error between input and local carrier

2.6 Calculating the position [11]

This section describes the basics for calculating a position by applying trilateration. First, the basic ideas of positioning are introduced, afterwards a short introduction on pseudoranges and how to measure them is given. The section will then describe how to determine an actual position on Earth by using several pseudoranges, what is the minimum number required and how to process when the system is overdetermined.

2.6.1 Positioning basics

The position of a certain point in space can be found from the distances measured from this point to some known positions in space. In Figure 2.12, the user position is on the x-axis; this is a one-dimensional case. Knowing both the satellite position S_1 and the distance to the satellite x_1, the user position can be either to the left or right of S_1. In order to determine the exact user position, the distance to another satellite with know position has to be measured. In this figure, the position of S_2 and x_2 uniquely determine the user position U.

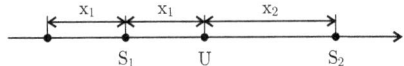

FIGURE 2.12: One-dimensional user position determination

Figure 2.13 shows a two-dimensional case. In order to determine the user position, three satellites and three distances are required. The trace of a point with constant distance to a fixed point is a circle in the two-dimensional case. As two circles intersect in two points, two satellites and two distances result in two possible solutions. Therefore, a third circle is needed to uniquely determine the user position.

The same approach can be applied if a three-dimensional case is considered. In this case, four satellites and four distances are needed to determine the position of the receiver. The equal-distance trace to a fixed point is a sphere in the three-dimensional case. The intersection of 2 spheres make a circle which intersects another sphere to produce two points. So, one more satellite is needed to determine which point is the user position.

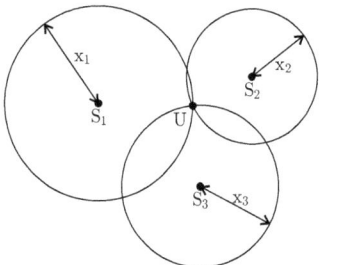

FIGURE 2.13: Two-dimensional user position determination

Application to navigation systems

The position of any GPS satellite k can be calculated from the transmitted ephemeris data, so the position S_k is known. The time from the satellite to the receiver can be measured by knowing when the message was sent by the satellite and when it was received by the receiver. The first parameter is transmitted in the GPS navigation data (see Section 2.2.3) and the second parameter can be measured in the receiver (with an ambiguity as described in Section 2.6.3). The relation between the time of flight and the distance can by made by the speed of light ($x = c \cdot t$), as the transmitted signals are electromagnetic waves that travel at the speed $c = 299'792.458$ km/s. So, the distance x_k between the satellite k and the receiver is also known. Now, if enough distances are available, the exact position of the receiver can be determined.

2.6.2 Basic equations for calculating the position

Figure 2.14 shows a constellation with three known points at locations r_1 or (x_1, y_1, z_1), r_2 or (x_2, y_2, z_2), and r_3 or (x_3, y_3, z_3), and an unknown point at r_u or (x_u, y_u, z_u). If the measured distances are accurate, the position of an unknown point r_u can be determined using the three distances p_1, p_2, and p_3, as given in Equation 2.17.

$$\begin{aligned} p_1 &= \sqrt{(x_1 - x_u)^2 + (y_1 - y_u)^2 + (z_1 - z_u)^2} \\ p_2 &= \sqrt{(x_2 - x_u)^2 + (y_2 - y_u)^2 + (z_2 - z_u)^2} \\ p_3 &= \sqrt{(x_3 - x_u)^2 + (y_3 - y_u)^2 + (z_3 - z_u)^2} \end{aligned} \quad (2.17)$$

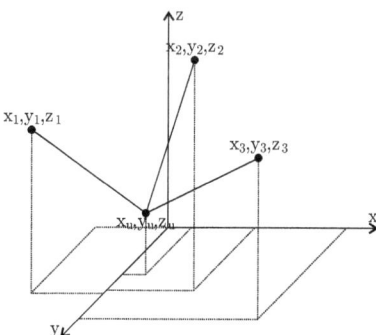

FIGURE 2.14: Use three known positions to find one unknown position

Because there are three unknown and three equations, the values of x_u, y_u, and z_u can be determined from these equations. As Equation 2.17 is of second order, there exist two possible solutions. However,

2.6.3 Measurement of pseudorange

Every satellite sends a signal at a certain time t_{si}. The receiver will receive the signal at a later time t_u. The distance between the receiver and the satellite i is given in Equation 2.18.

$$\rho_{iT} = c \cdot (t_u - t_{si}) \tag{2.18}$$

where c is the speed of light, p_{iT} is often referred to as the true value of pseudorange from the user to the satellite i, t_{si} is referred to as the true time of transmission from satellite i, and t_u is the true time of reception.

From a practical point of view, it is difficult, if not impossible, to obtain the correct time from the satellite or the user. The actual satellite clock time t'_{si} and the actual user clock time t'_u are related to the true time as given in Equation 2.19.

$$\begin{aligned} t'_{si} &= t_{si} + \Delta b_i \\ t'_u &= t_u + b_{ut} \end{aligned} \tag{2.19}$$

where Δb_i is the satellite clock error and b_{ut} is the user clock error.

The measured distance is called *pseudorange* ρ because it is the range determined by multiplying the signal propagation velocity c (speed of light) with the time difference between two non synchronized clocks (the satellite clock and the receiver clock). The measurement contains:

1. The geometric satellite-to-user range;
2. An offset attributed to the difference between system time and the user clock;
3. An offset between system time and satellite clock.

Besides the clock errors, there are other factors affecting the pseudorange measurements as given in Equation 2.20.

$$\rho_i = \rho_{iT} + \Delta D_i - c \cdot (\Delta b_i - b_{ut}) + c \cdot (\Delta T_i + \Delta I_i + v_i + \Delta v_i) \tag{2.20}$$

where ΔD_i is the satellite position error effect on the range, ΔT_i is the tropospheric delay error, ΔI_i is the ionospheric delay error, v_i is the receiver measurement noise error, and Δv_i is the relativistic time correction.

If not corrected, these errors will nevertheless cause an inaccuracy on the user position. As the user clock error cannot be corrected through the received information, it will remain as an unknown. As a

Chapter 2 Introduction to GPS receivers 33

result, Equation 2.17 has to be modified as given in Equation 2.21.

$$\begin{aligned}
\rho_1 &= \sqrt{(x_1 - x_u)^2 + (y_1 - y_u)^2 + (z_1 - z_u)^2} + b_u \\
\rho_2 &= \sqrt{(x_2 - x_u)^2 + (y_2 - y_u)^2 + (z_2 - z_u)^2} + b_u \\
\rho_3 &= \sqrt{(x_3 - x_u)^2 + (y_3 - y_u)^2 + (z_3 - z_u)^2} + b_u \\
\rho_4 &= \sqrt{(x_4 - x_u)^2 + (y_4 - y_u)^2 + (z_4 - z_u)^2} + b_u
\end{aligned} \quad (2.21)$$

where b_u is the user clock bias error expressed in distance, which is related to the quantity b_{ut} by $b_u = c \cdot b_{ut}$. In Equation 2.21, four equations are needed to solve for four unknowns x_u, y_u, z_u, and b_u. Thus, in a GPS receiver, a minimum of four satellites is required to solve for the three-dimensional user position.

2.6.4 User position from pseudoranges

One common way to solve Equation 2.21 is to linearize them. The above equations can be written in a simplified form as given in Equation 2.22.

$$\rho_i = \sqrt{(x_i - x_u)^2 + (y_i - y_u)^2 + (z_i - z_u)^2} + b_u \quad (2.22)$$

where $i = 1, 2, 3, 4$, and x_u, y_u, z_u, and b_u are the unknowns. The pseudorange ρ_i and the positions of the satellites x_i, y_i, z_i are known.

Differentiate Equation 2.22 leads to Equation 2.23.

$$\begin{aligned}
\delta\rho_i &= \frac{(x_i - x_u) \cdot \delta x_u + (y_i - y_u) \cdot \delta y_u + (z_i - z_u) \cdot \delta z_u}{\sqrt{(x_i - x_u)^2 + (y_i - y_u)^2 + (z_i - z_u)^2}} + \delta b_u \\
&= \frac{(x_i - x_u) \cdot \delta x_u + (y_i - y_u) \cdot \delta y_u + (z_i - z_u) \cdot \delta z_u}{\rho_i - b_u} + \delta b_u
\end{aligned} \quad (2.23)$$

where δx_u, δy_u, δz_u, and δb_u can be considered as the only unknowns. The quantities x_u, y_u, z_u, and b_u are treated as known values with assumed initial values. From these initial set of values, a new set of δx_u, δy_u, δz_u, and δb_u can be calculated. These values are used to modify the original x_u, y_u, z_u, and b_u to find another new set of solutions which can be considered again as known quantities. This process continues until the absolute values of δx_u, δy_u, δz_u, and δb_u are very small and within a certain predetermined limit. The final values of x_u, y_u, z_u, and b_u are the desired solution. This method is often referred to as the *iteration method*.

With δx_u, δy_u, δz_u, and δb_u as unknowns, the above equation becomes a set of linear equations. This procedure is often referred to as linearization. The above equation can be written in matrix form as given in Equation 2.24.

$$\begin{bmatrix} \delta\rho_1 \\ \delta\rho_2 \\ \delta\rho_3 \\ \delta\rho_4 \end{bmatrix} = \begin{bmatrix} \alpha_{11} & \alpha_{12} & \alpha_{13} & 1 \\ \alpha_{21} & \alpha_{22} & \alpha_{23} & 1 \\ \alpha_{31} & \alpha_{32} & \alpha_{33} & 1 \\ \alpha_{41} & \alpha_{42} & \alpha_{43} & 1 \end{bmatrix} \cdot \begin{bmatrix} \delta x_u \\ \delta y_u \\ \delta z_u \\ \delta b_u \end{bmatrix} \quad (2.24)$$

where

$$\alpha_{i1} = \frac{x_i - x_u}{\rho_i - b_u}; \quad \alpha_{i2} = \frac{y_i - y_u}{\rho_i - b_u}; \quad \alpha_{i3} = \frac{z_i - z_u}{\rho_i - b_u}. \quad (2.25)$$

The solution of Equation 2.24 is given in Equation 2.26.

$$\begin{bmatrix} \delta x_u \\ \delta y_u \\ \delta z_u \\ \delta b_u \end{bmatrix} = \begin{bmatrix} \alpha_{11} & \alpha_{12} & \alpha_{13} & 1 \\ \alpha_{21} & \alpha_{22} & \alpha_{23} & 1 \\ \alpha_{31} & \alpha_{32} & \alpha_{33} & 1 \\ \alpha_{41} & \alpha_{42} & \alpha_{43} & 1 \end{bmatrix}^{-1} \cdot \begin{bmatrix} \delta\rho_1 \\ \delta\rho_2 \\ \delta\rho_3 \\ \delta\rho_4 \end{bmatrix} \quad (2.26)$$

where $[\ ]^{-1}$ represents the inverse of the α matrix. This equation obviously does not provide the needed solution directly but it can be obtained from it. In order to find the desired position solution, this equation must be used repetitively in an iterative way. To determine whether the desired result is reached, a quantity (called *Geometric Dilution Of Precision (GDOP)*) can be defined as given in Equation 2.27.

$$\text{GDOP} = \sqrt{\delta x_u^2 + \delta y_u^2 + \delta z_u^2 + \delta b_u^2} \quad (2.27)$$

When this value is less than a certain predetermined threshold, the iteration will stop and the position is said to be "determined". Sometimes, the clock bias b_u is not included in Equation 2.27.

2.6.5 Position with more than four satellites

When more than four satellites are available, a more popular approach to solve the user position is to use all the satellites so the position solution can be obtained in a similar way. If there are n satellites available with $n > 4$, Equation 2.22 can be written as given in Equation 2.28.

$$\rho_i = \sqrt{(x_i - x_u)^2 + (y_i - y_u)^2 + (z_i - z_u)^2} + b_u \quad (2.28)$$

where $i = 1, 2, 3, \ldots, n$. The only difference between this equation and Equation 2.22 is that $n > 4$.

Linearizing Equation 2.28 results in Equation 2.29.

$$\begin{bmatrix} \delta\rho_1 \\ \delta\rho_2 \\ \delta\rho_3 \\ \delta\rho_4 \\ \vdots \\ \delta\rho_n \end{bmatrix} = \begin{bmatrix} \alpha_{11} & \alpha_{12} & \alpha_{13} & 1 \\ \alpha_{21} & \alpha_{22} & \alpha_{23} & 1 \\ \alpha_{31} & \alpha_{32} & \alpha_{33} & 1 \\ \alpha_{41} & \alpha_{42} & \alpha_{43} & 1 \\ \vdots & \vdots & \vdots & \vdots \\ \alpha_{n1} & \alpha_{n2} & \alpha_{n3} & 1 \end{bmatrix} \cdot \begin{bmatrix} \delta x_u \\ \delta y_u \\ \delta z_u \\ \delta b_u \end{bmatrix} \qquad (2.29)$$

where

$$\alpha_{i1} = \frac{x_i - x_u}{\rho_i - b_u}; \qquad \alpha_{i2} = \frac{y_i - y_u}{\rho_i - b_u}; \qquad \alpha_{i3} = \frac{z_i - z_u}{\rho_i - b_u}. \qquad (2.30)$$

Equation 2.29 can be written in a simplified form as given in Equation 2.31.

$$\delta\rho = \alpha \cdot \delta x \qquad (2.31)$$

where $\delta\rho$ and δx are vectors and α is a matrix.

They can be written as given in Equation 2.32.

$$\delta\rho = \begin{bmatrix} \delta\rho_1 & \delta\rho_2 & \cdots & \delta\rho_n \end{bmatrix}^T$$
$$\delta x = \begin{bmatrix} \delta x_u & \delta y_u & \delta z_u & \delta b_u \end{bmatrix}^T$$
$$\alpha = \begin{bmatrix} \alpha_{11} & \alpha_{12} & \alpha_{13} & 1 \\ \alpha_{21} & \alpha_{22} & \alpha_{23} & 1 \\ \alpha_{31} & \alpha_{32} & \alpha_{33} & 1 \\ \alpha_{41} & \alpha_{42} & \alpha_{43} & 1 \\ \vdots & \vdots & \vdots & \vdots \\ \alpha_{n1} & \alpha_{n2} & \alpha_{n3} & 1 \end{bmatrix} \qquad (2.32)$$

where $[\ ]^T$ represents the transpose of the matrix. Since α is not a square matrix, it cannot be inverted directly, but Equation 2.29 is still a linear equation. If there are more equations than unknowns in a set of linear equations, the least-squares approach can be used to find the solution. The pseudoinverse of the α matrix can be used to obtain the solution that is

$$\delta x = [\alpha^T \cdot \alpha]^{-1} \cdot \alpha^T \cdot \delta \rho \tag{2.33}$$

From this equation, the values of δx_u, δy_u, δz_u, and δb_u can be found. In general, the least-squares approach produces a better solution than the one obtained from four satellites because more data is used.

2.7 Summary

This chapter gave an overview of the GPS signal characteristics and the different processing blocks of a generic GNSS receiver, including the tracking loops and the navigation solution.

The transmitted signals can only be recovered thanks to the spread-spectrum modulation as the received signal power level on Earth lies below the thermal noise floor. Thereto, every satellite transmits its own spreading sequence (called pseudorandom noise (PRN) code) which makes it possible to find the corresponding satellite in the picked up signal by correlating the received signal with the corresponding locally generated PRN code. If the satellite is present in the signal and if the PRN codes are perfectly aligned, a peak can be detected in the correlation output.

Three different acquisition architectures, processing the signal either in the time or in the frequency domain, have been presented. The acquisition stage determines – for every satellite – the beginning of the PRN code in the incoming signal and the Doppler frequency shift caused by the motion of the satellite and the receiver itself. The transmitted carrier frequency can be shifted by up to $\pm$ 10 kHz and this offset has to be determined precisely to remove any residual frequency component on the signal.

The next section explained the tracking stage, starting with the description of a first and second order PLL. Afterwards, the architecture for the code and carrier tracking is presented together with the corresponding discriminators.

The chapter is closed with the description of the navigation solution that showed that at least four satellites are required to determine a valid 3D-position (the forth satellite is required due of the unknown time).

Chapter 3

Software receivers

3.1 Introduction

This chapter provides a definition of the term *Software Receiver (SR)* and describes some of the challenges and the current status of real-time software receivers. It starts with a short history and then gives an overview of the importance and the impact of this concept, i.e. what are the main advantages of software receivers and where can they be used? Afterwards, some of the challenges are explained, as well as an estimation of the needed processing power. The last section covers the existing solutions, mainly the algorithm side, i.e., how are current software receivers implemented and how do they overcome the challenges presented before?

3.2 History

Until the late 1980s, almost all radio designs were based on hardware technology. This is mainly because it was the only technology able of handling the requirements in terms of processing power and speed. In the early 1990s, the U.S. military services were facing several communication-related challenges such as ensuring communication with current allies, offering a global support structure, and controlling the (often tremendous) costs of R&D and purchasing. At that time, military radio designs were optimized for a single, specific field application and typically designed for a 30-year development life span. During the 1990s, commercial applications started to drive the global technology development so that the effective lifetime of a commercial component fell to less than two years.

Influenced by this change in equipment design and development environment, a U.S. Department of Defense (DoD) project named 'Speakeasy' was undertaken with the objective of finding and proving a concept of a programmable waveform, multiband, multimode radio [20]. The Speakeasy project demonstrated the approach that is valid for most software receivers: the Analog-to-Digital Converter

(ADC) is placed as close as possible to the antenna and all baseband functions that receive digitized IF data input are processed in a programmable microprocessor using software techniques rather than hardware elements, such as correlators.

Until the late 1990s, due to the limited processing power of available microprocessors, the most time and resource consuming signal operations (like the correlation) could only be practically implemented in hardware. However, with the development and availability of new and more powerful CPUs and DSPs and new algorithms, it became possible to implement (some of) these operations in software.

In 1990, researchers at the NASA/Caltech Jet Propulsion Laboratory introduced a signal acquisition technique for CDMA systems that was based on the FFT. This technique was extended by the work of researchers at the Technical University of Delft using FFT and inverse-FFT-based signal acquisition for GPS [16], [21]. Since then, this method has been widely adopted in GNSS software receivers because of its simplicity and efficiency of processing load.

In 1996, researchers at Ohio University provided a direct digitization technique – called the bandpass sampling technique – that allowed the placing of the ADCs closer to the RF portions of GNSS software receivers. Until this time, the implemented software receiver or software-defined receivers (SDR) in university laboratories post-processed the data due to the lack of processing power mentioned earlier.

However, the GNSS SR boom really started with the development of real-time processing capability, mainly with the Ph.D. thesis of D. Akos [22]. In that work, the core concepts were implemented and 30 seconds of GPS data were processed until a position fix was achieved. To demonstrate the flexibility, GLONASS signals were successfully acquired and tracked. The work by Akos performed all the signal processing in post-processing mode, but soon thereafter, a real-time implementation was achieved on a DSP and on a general-purpose PC. This was first accomplished on a digital signal processor (DSP) and later on a commercial conventional personal computer (PC) [23]. Today, the DSPs are more and more replaced by specialized processors for embedded applications.

Another precursor in the field of SR was Philip Mattos who published already in 1989 an article about a "low-cost hand-held GPS navigation system receiver" where he implemented the correlation process on a transputer (*transistor computer*) developed by Inmos [24].

3.3 Importance of software receivers

The idea of a software receiver is to realize the data processing blocks, implemented on traditional receivers in hardware, in software and to sample the analog input signal as close to the antenna as possible. Thus, the hardware is reduced to the minimum (e.g., antenna and analog to digital converters) while all the signal processing is done in software. As current mobile devices (such as personal digital assistants and smartphones) include more and more computing power and system features it becomes possible to integrate a complete GNSS receiver with very few external components.

One advantage of a software receiver lies clearly in the possibility to realize a low cost solution as the – already available – system resources, such as the calculation power and system memory, can be employed and a specific GNSS baseband chip can be avoided. The receiver can interact more flexibly with other software running on the same system. Provision of aiding data via the mobile phone data link, for example, can be more easily achieved. Also, integration with other sensors (e.g., WLAN) or any other user software provides potential advantages in obtaining an integrated position solution.

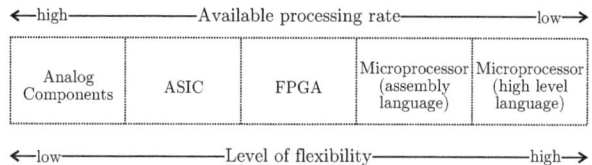

FIGURE 3.1: Processing rate versus level of flexibility

Another advantage resides in the flexibility (as shown in Figure 3.1) for adapting to new signals and frequencies. Indeed, an update can easily be performed by changing some parameters and algorithms in software while it would require a complete re-development for a standard hardware receiver.

Updating capabilities may become even more important in the future as the world of satellite navigation is in complete effervescence [25]:

o Europe is developing its own solution (named Galileo) that is foreseen to be operational in early 2014 [7];
o China is about to undertake a fundamental re-development of its current navigation system (named Compass);
o Russia is investing a huge amount of money in its GLONASS system to bring it back to full operation;
o The U.S. GPS system will see some fundamental improvements during the next few years with new frequencies and new modulation techniques.

At the same time, augmentation systems (either space based or land based) will be developed all over the world. These future developments will increase the number of accessible satellites available to every user, with the advantage of better coverage and higher accuracy. However, to take full advantage of the new satellite constellations and signals, new GNSS receivers and algorithms must be developed that can easily be adopted to the new specifications of the signals and the modulation types. Therefore, software receivers will gain high interest during the next years.

The disadvantages of a software solution compared to a hardware solution include increased processing load and power consumption. Today, the processing load necessary to track one satellite signal varies between 3 and 20 MIPS in commercially available products, depending on the power of the received signal [13].

3.4 Definition and types

The definition of a SR always brings some confusion among researchers and engineers in the field of communications and GNSS. For example, a receiver containing multiple hardware parts which can be reconfigured by setting a software flag or a hardware pin of the chipset is regarded by some communication engineers to be a SR (although the term *Software-Defined Radio (SDR)* or even *Parameter-controlled SDR (PaC-SDR)* is more appropriate). In this document, however, only the widely accepted SR definition in the field of GNSS is considered, that is: a receiver in which all the baseband signal processing is performed in software by a programmable unit (processor). A SR is not tailored to a specific chip or platform, and it is therefore possible to reuse its code across different underlying architectures.

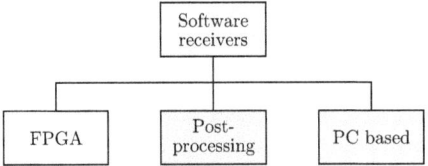

FIGURE 3.2: Software receiver types

Nowadays, software receivers can be grouped in three main categories as shown in Figure 3.2:

o The first category regroups all the receivers that are based on FPGA. These receivers are sometimes also referred to the domain of SR as they can be reconfigured in the field by software or because a CPU can be implemented directly into the FPGA (such as the NIOS II processor [26]).
o The second category, post-processing receivers includes, among others, the countless software tools or lines of code for testing new algorithms and for analyzing the GNSS signal, for example, to investigate GPS satellite failure or to decrypt unpublished codes.
o Finally, the third category is the PC based real-time capable SR group. This category includes the receivers that perform a real-time operation in software, running on a standard computer or an embedded system.

Only the last category (real-time PC based software receivers) will be further considered in this document.

3.5 An ideal software receiver

A rather extensive definition was given by Joseph Mitola who coined the term *Software Radio* in 1992. Although he is talking about a general radio (i.e., not only for the satellite communication frequency bands), his definition can be very well adopted for GNSS software receivers.

> A software radio is a radio whose channel modulation waveforms are defined in software. [...] The receiver employs a wideband ADC that captures all of the channels of the software radio node. The receiver then extracts, downconverts and demodulates the channel waveform using software on a general purpose processor. Software radios employ a combination of techniques that include multi-band antennas and RF conversion; wideband ADC; and the implementation of IF, baseband and bitstream processing functions in general purpose programmable processors. The resulting software radio in part extends the evolution of programmable hardware, increasing flexibility via increased programmability. [27]

This means, that instead of using analog circuits or a specialized Digital Signal Processor (DSP) to process the radio signal, the digitized signal is processed by an architecture independent high level software running on a general purpose processor.

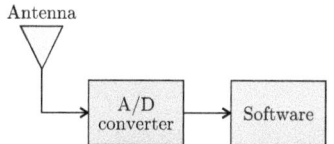

FIGURE 3.3: Scheme of ideal software receiver

In the ideal case, the only hardware needed beside a computer is an antenna and an analog-to-digital converter. A SR would thus look as depicted in Figure 3.3. The transmitted radio signal is picked up by an antenna and then fed into an ADC to sample it. Once digitized, the signal is sent to some general purpose computer (e.g., an embedded PC) for processing.

3.6 Challenges

This section describes some of the challenges when realizing the ideal software receiver (as seen in Section 3.5) or when a classical hardware approach is directly converted to software. The discussion will include the following aspects:

1. Data rate;

2. Signal sample conversion;
3. Real-time baseband processing.

For each point, a brief estimation of the needed processing power is given that shows why this aspect can be considered as a challenge. A more detailed calculation of the different blocks of a software receiver can be found in Section 3.7.

3.6.1 Data rate

The ideal software receiver would place the ADC as close as possible to the antenna in order to reduce the hardware parts to the minimum (see Figure 3.3). In that sense, the most straightforward approach consists in digitizing the data directly at the antenna without pre-filtering or pre-processing. But as the Nyquist theorem must be fulfilled, this translates into a data rate that is – for the time being – too high to be processed by a microprocessor.

Considering the GPS L1 signal and assuming 1 quantization bit per sample, this leads to the following values:
$$F_{GPS_{L1}} = 1.575 \text{ GHz}$$
$$F_s \geq 2 \cdot F_{GPS_{L1}} = 3.15 \text{ GHz} \tag{3.1}$$
$$\text{Data rate} \geq 3.15 \text{ GBit/s} \approx 394 \text{ MB/s}$$

This data rate is relatively difficult to handle even for today's available computer systems. Not only the interface between the RF front-end and the computer has to support this speed, but the computer must also be able to handle the data in real-time. Another exigence arises if the data has to be saved locally (either for post-processing or for later data analysis). It is important to note that the number of quantization bits is directly proportional to the data rate (doubling the number of quantization bits also doubles the data rate).

In order to reduce the resulting data rate, a solution such as a low intermediate frequency (low IF) or a sub-sampling analog front-end can be chosen.

In a low IF front-end, the incoming signal is down-converted to a lower intermediate frequency of several megahertz. This allows working with a sampling (and a resulting data) rate that can be more easily handled by a microprocessor.

The sub-sampling technique exploits the fact that the effective signal bandwidth is much lower than the carrier frequency ($BW_{GPS_{L1}} = 2$ MHz $\leftrightarrow F_{GPS_{L1}} = 1.575$ GHz). Therefore, not the carrier frequency but the signal bandwidth of the GNSS signal must be respected by the Nyquist theorem (assuming appropriate band-pass filtering). In this case, the modulated signal is under-sampled to achieve frequency translation via intentional aliasing. Again, if the GPS L1 signal is taken as an example with 1 quantization bit per sample, this leads to the following values:

$$\mathrm{BW}_{GPS_{L1}} = 2 \text{ MHz}$$
$$F_s \geq 2 \cdot \mathrm{BW}_{GPS_{L1}} = 4 \text{ MHz} \quad (3.2)$$
$$\text{Data rate} \geq 4 \text{ MBit/s} \approx 394 \text{ kB/s}$$

However, as the sub-sampling approach causes a degradation of the C/N_0 value (see Section 2.3) and is still difficult to implement due to current hardware and resources limitations (e.g., $\mathrm{BW}_{ADC} \geq 2 \cdot \mathrm{BW}_{GPS_{L1}}$), a more classical solution based on an analog IF down-conversion is often used. That means that the signal is first down-converted to an intermediate frequency of several megahertz and afterwards digitized, resulting in a slightly higher data rate than for the sub-sampling technique. However, the sub-sampling approach is currently often used for bringing the signal at IF to baseband.

3.6.2 Signal sample conversion

Navigation signals within a software receiver are usually represented by integer samples of a given bit size. The term "signal" refers to both the received navigation signal(s) and the internally generated signal(s). Usually, a low number of bits is sufficient to represent those signals. Representing a GNSS signal with 3 bits causes a loss of 0.2 dB (in the absence of interference); internally generated sine/cosine carriers can be represented by a 1 bit amplitude causing a loss of 0.9 dB [13].

Two hardware-related factors are present that determine more rigorously how many bits shall be used to represent the signals. They are:

1. The number of bits provided by the ADC or the maximum bandwidth between the ADC and the PC;
2. The optimal integer format supported by the CPU.

Consequently, a conversion of the input ADC bits to the best internal working format of the CPU is generally needed. T. Pany made in his latest book [13] a calculation of the needed processing power for the required bit conversion operation. He based his algorithm on a lookup table method where the input value (of a given bit size and format) is used as an index of a precomputed table, and the table entries represent the output values (of a different bit size and format). Using a lookup table method avoids arithmetic calculations that are otherwise required for the conversion. One disadvantage of this method is that it requires frequent memory accesses.

The numerical performance was evaluated by converting 2 bit input values into 16 bit output values. The input stream contains in each byte two 2 bit values in the LSB positions. The two 2 bit values are converted into two 16 bit values in one step and the 16-values are represented as one 32 bit value. The code is written directly in the assembler language and one loop cycle (i.e., converting two values) takes

10.96 clock ticks on a test system. The system is running at a CPU clock speed of 2.26 GHz, thus 412 Msamples can be converted per second. This value is compared to the needs of a triple-frequency GNSS software receiver with a sampling frequency of 40.96 MHz. It was found that the conversion (which is a more or less trivial operation) consumes 28% of the maximum CPU load on the test system. It was also demonstrated that the most time consuming (or clock-tick consuming) instruction is the lookup operation, taking 7.81 clock ticks on average (on a total of 10.96 clock ticks).

It is almost impossible to optimize the code. One small possibility is to keep the lookup table itself in the CPU's L1 cache, but the input and output streams have – nevertheless – to be read/written from/to the main memory.

3.6.3 Baseband processing

Real-time carrier generation and mixing

In hardware receivers, the local carrier replicas are normally generated in real-time by the means of a NCO which performs the role of a digital waveform generator by incrementing an accumulator by a per-sample phase increment. The resulting value is then converted to the corresponding amplitude value in order to recreate the waveform at any desired phase offset. The frequency resolution of a 32 bit accumulator is typically in the range of a few millihertz and the sampling frequency in the range of a few megahertz. Assuming that a lookup table (LUT) address can be obtained with 2 logical operations (one shift and one mask) and the corresponding LUT value read with 1 memory access – which is quite optimistic – the amount of needed operations to generate and mix the complex waveforms in a complex IF architecture is given in Table 3.1. The exact calculations of these values can be found in Section 3.7.2.

Total integer additions	$3 \cdot N_{ch} \cdot F_s \cdot T_{int}$
Total integer multiplications	$4 \cdot N_{ch} \cdot F_s \cdot T_{int}$
Total logical operations	$3 \cdot N_{ch} \cdot F_s \cdot T_{int}$

TABLE 3.1: Number of required operations for carrier generation and mixing in a complex IF architecture

On an older Intel Pentium 4 processor an integer addition takes one clock cycle, while an integer multiplication takes 14 clock cycles [28]. If the equations above are transformed into numerical values (assuming a 12-channel GPS receiver with an integration time of 1 ms and a sampling frequency of 4 MHz), a total number of $2.83 \cdot 10^9$ clock cycles is obtained per second (only integer additions and multiplications). This corresponds to a CPU running at roughly 2.9 GHz – considering only the operations for generating and mixing the carrier. This value clearly shows that a real-time generation in software is not possible in the same way as in a hardware implementation.

Correlation throughput

The correlators are probably the most important part of a GNSS receiver. The correlation algorithm performs three tasks:

1. Down-converting the signal affected by a Doppler shift to baseband;
2. Multiplying the baseband signals with locally generated PRN sequences;
3. Accumulating the result over a given integration time.

G. Heckler made an estimation of the needed processing power to perform the real-time correlation for a 12-channel receiver [29]. He considers a 12-channel receiver with 3 correlators per channel (early, prompt, and late) and an integration time of 1 millisecond. These parameters give a required correlation bandwidth of the software receiver of 36'000 correlations/second to achieve real-time performance. In order to retain the responsiveness of the PC running a modern multi-tasking OS, the software receiver should be limited to 50% CPU load. Thus the microprocessor must be able to achieve 72'000 correlations/second in an idealized benchmark.

He used a processor purchased in 2006 that runs at 2.0 GHz, a rate of $2 \cdot 10^9$ clock cycles per second. Such a processor could allot 27'777 clock cycles to achieve a single correlation [29]. This value was assumed to be a practical upper bound on the number of clock cycles a correlation can require. Today, this value would be smaller as the internal architecture of the processor was optimized during the last 4 years. But nevertheless, the following considerations stay the same and the final conclusion will not be very different.

To capture the complete main lobe of the C/A code signal, a minimum sample rate of $2.046 \cdot 10^6$ samples/second (for complex value) or $4.092 \cdot 10^6$ samples/second (for real values) would be required. Thus, the C/A code correlation will operate on a vector, **IF**, of 2'046 or 4'096 samples. G. Heckler took the example of real data, so the vector is 4'096 samples long. To produce a complex correlation for real data, the algorithm is the following:

1. Generate the in-phase carrier removal vector, **C**;
2. Generate the quadrature carrier removal vector, **S**;
3. Generate the C/A code replica, **PRN**;
4. Multiply the **IF** by the **PRN** vector;
5. Multiply **IF·PRN** by **C** and accumulate to generate the in-phase correlation;
6. Multiply **IF·PRN** by **S** and accumulate to generate the quadrature correlation.

If the computational cost of the generation of the carrier removal vectors and the C/A code replica are ignored (point 1 – 3), it takes three multiplications and two additions, per sample, to generate the complex correlation. For the entire 4'092 sample vector, the approximate number of operations is 12'000 multiplications and 8'000 additions.

The cycle count for the 4'092 sample vector is approximately 130'000 versus the 27'777 cycle limit established previously (taking the same number of required clock cycles for the two operations for a Pentium 4 CPU as above). An initial look at the correlation algorithm clearly indicates that a real-time operation of 12 correlators is not possible.

3.7 Existing solutions and their complexity

This section describes the needed processing power of different architectures and algorithms in a GNSS receiver. The first part will analyze the number of integer operations needed for the correlation in a real and in a complex IF architecture. Then the different acquisition methods are analyzed and compared in terms of needed operations and finally, an estimation of the FFT performance is given and the term *number of equivalent correlators* will be explained.

3.7.1 Baseband processing architecture

The choice of the baseband processing architecture has an important impact on the needed computational power. The advantages of choosing either the real or the complex IF architecture will not be discussed in this document, but can be found in [11].

Real IF architecture

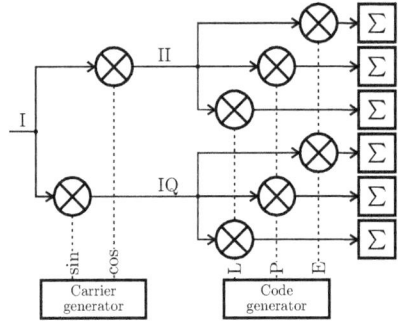

FIGURE 3.4: Real IF baseband processing architecture

In a real IF baseband architecture, the incoming signal is only represented by the in-phase part (referenced hereafter to as I) as depicted in Figure 3.4.

The mathematical description of the II and IQ signal branches for the carrier and Doppler removal can be found in Appendix C.

Chapter 3 Software receivers

The amount of integer operations necessary to process the real baseband signal for N_{ch} channels (without the carrier and code generation) is given in Table 3.2. These equations can be obtained directly by analyzing Figure 3.4.

	Additions	Multiplications
Carrier mixing	0	$2 \cdot N_{ch} \cdot F_s \cdot T_{int}$
Code mixing	0	$2 \cdot N_{ch} \cdot N_{cor} \cdot F_s \cdot T_{int}$
Accumulation	$2 \cdot N_{ch} \cdot N_{cor} \cdot F_s \cdot T_{int}$	0
Total	$2 \cdot N_{ch} \cdot N_{cor} \cdot F_s \cdot T_{int}$	$2 \cdot N_{ch} \cdot F_s \cdot T_{int} \cdot (N_{cor} + 1)$

TABLE 3.2: Number of required operations for a real IF architecture

Complex IF architecture

In a complex IF baseband architecture, the incoming signal is complex (quadrature part hereafter denoted as Q) as illustrated in Figure 3.5.

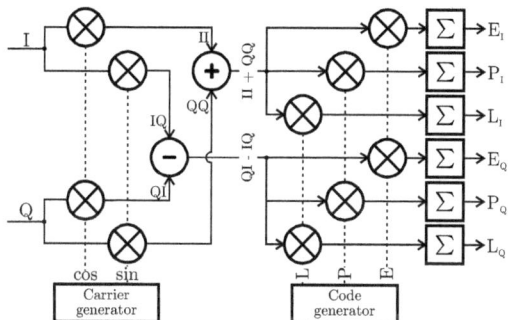

FIGURE 3.5: Complex IF baseband processing architecture

Since there are two separate input channels, the Nyquist condition becomes $F_{s_{IQ}} = F_s/2 > BW$, thus the sampling frequency can be divided by two without loosing any information.

The mathematical description of the different signal branches and the corresponding recombinations for the carrier and Doppler removal can be found in Appendix C.

The incoming signal is sequentially processed at the system sampling frequency F_s for:

o Residual carrier (Doppler) removal;
o PRN code removal;
o Integration and dump.

The amount of integer operations necessary to process the complex baseband signal (without the carrier and the code generation) is given in Table 3.3.

	Additions	Multiplications
Carrier mixing	0	$4 \cdot N_{ch} \cdot F_{s_{IQ}} \cdot T_{int}$
I & Q recomb.	$2 \cdot N_{ch} \cdot F_{s_{IQ}} \cdot T_{int}$	0
Code mixing	0	$2 \cdot N_{ch} \cdot N_{cor} \cdot F_{s_{IQ}} \cdot T_{int}$
Accumulation	$2 \cdot N_{ch} \cdot N_{cor} \cdot F_{s_{IQ}} \cdot T_{int}$	0
Total	$N_{ch} \cdot F_s \cdot T_{int} \cdot (N_{cor} + 1)$	$N_{ch} \cdot F_s \cdot T_{int} \cdot (N_{cor} + 2)$

TABLE 3.3: Number of required operations for a complex IF architecture

Comparison

Considering the configuration with $N_{ch} = 12$, $N_{cor} = 3$, $T_{int} = 1$ ms and $F_s = 4$ MHz, the numerical values for the number of needed operations of the two architectures is given in Table 3.4.

	Additions	Multiplications
Real IF architecture	$2.88 \cdot 10^5$	$3.84 \cdot 10^5$
Complex IF architecture	$1.92 \cdot 10^5$	$2.40 \cdot 10^5$

TABLE 3.4: Comparison of numerical values of required operations for a real and a complex IF baseband architecture

Although the complex IF baseband requires additional adders and mixers, the global amount of operations is reduced (by 30% in the given example). This is due to the reduced sampling frequency. In terms of performance requirements and complexity, the complex IF baseband architecture is well suited for a software receiver implementation.

3.7.2 Carrier generation and mixing algorithms

The local generation of a carrier replica is necessary to be able to perform the Doppler frequency removal. The trigonometric functions available in many compilers and the Taylor series decomposition are too inefficient for a real-time sine and cosine computation and can therefore not be used for the carrier generation in a software receiver.

Real-time generation and mixing with integer arithmetic

In hardware receivers, the real-time carrier generation is generally achieved by the means of a NCO. The process is illustrated in Figure 3.6.

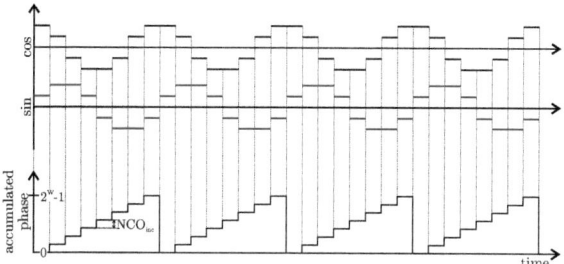

FIGURE 3.6: NCO generating a 2 bit quantized waveform

The generated frequency F_{car} depends on the accumulator bit-width W and is proportional to the sampling clock F_s and the phase increment NCO_{inc}, as given in Equation 3.3.

$$F_{car} = \frac{F_s \cdot NCO_{inc}}{2^W} \qquad (3.3)$$

The NCO resolution increases with the width W of the accumulator and is given by Equation 3.4.

$$\Delta F_{car} = \frac{1}{2} \cdot \left[\frac{F_s \cdot (NCO_{inc}+1)}{2^W} - \frac{F_s \cdot NCO_{inc}}{2^W} \right] = \frac{F_s}{2^{W+1}} \qquad (3.4)$$

The frequency resolution is typically in the range of a few millihertz for a 32 bit accumulator and a sampling frequency of a few megahertz.

Assuming that a LUT address can be obtained with 2 logical operations (1 shift and 1 mask) and the corresponding LUT value with 1 memory access – which is quite optimistic – the amount of operations needed for the generation and the mixing of the carrier in a complex IF architecture is given in Table 3.5. The number of logical operations corresponds to $3 \cdot N_{ch} \cdot F_s \cdot T_{int}$ and is not included in the table for a clearer view.

	Additions	Multiplications
Carrier generation	$1 \cdot N_{ch} \cdot F_s \cdot T_{int}$	0
Carrier mixing	0	$4 \cdot N_{ch} \cdot F_s \cdot T_{int}$
Carrier recombination	$2 \cdot N_{ch} \cdot F_s \cdot T_{int}$	0
Total	$3 \cdot N_{ch} \cdot F_s \cdot T_{int}$	$4 \cdot N_{ch} \cdot F_s \cdot T_{int}$

TABLE 3.5: Number of required operations for carrier generation and mixing with integer arithmetic

Considering the configuration with $N_{ch} = 12$, $T_{int} = 1$ ms and $F_s = 4$ MHz, the numerical values for the number of needed operations of the real-time carrier generation with integer arithmetic is given in Table 3.6.

	Additions	Multiplications	Logical operations
Real-time carrier generation and mixing	$1.44 \cdot 10^5$	$1.92 \cdot 10^5$	$1.44 \cdot 10^5$

TABLE 3.6: Numerical values of required operations for real-time carrier generation and mixing with integer arithmetic

The real-time carrier generation and mixing is computationally expensive and is consequently not implemented this way in real-time software receivers. However, Single Instruction Multiple Data (SIMD) operations (see Section 3.8.1) can be advantageously used to parallelize the computations and speed up the processing time. For example, the phase accumulation and the carrier mixing can be performed concurrently for several channels.

Lookup method I

This method was first introduced by Ledvina in [30]. He pre-computed different carrier frequencies at the desired Doppler frequencies and stored them in a lookup table. As it would require several gigabytes of memory to store all the possible frequencies, the values are recorded on a coarse frequency grid with zero phase and at the RF front-end sampling frequency (i.e., in an over-sampled representation). The limited number of available carrier frequencies introduces a supplementary mismatch δf in the Doppler frequency removal process which causes a correlation loss. In order to recreate an estimation of what would have been computed with the correct frequency and phase, the accumulation results are rotated properly to δf.

A decrease of the C/N_0 value is expected from using an inexact frequency. The worst-case decrease is expressed as a function of the frequency grid spacing Δf and is given by Equation 3.5 [31].

$$\Delta \text{SNR} = 20 \cdot \log_{10} \left(\frac{\sin(\pi \cdot \Delta f \cdot T_{int})}{\pi \cdot \Delta f \cdot T_{int}} \right) \tag{3.5}$$

Thus, a Δf of 175 Hz causes a worst-case SNR loss of 0.44 dB for $T_{int} = 1$ ms.

Assuming a Doppler frequency range of ± 10 kHz, 134 different frequency bins with a resolution of 150 Hz, and a sampling frequency of 4 MHz, the memory requirements are given in Table 3.7.

To keep the SNR loss constant, longer coherent integration time requires a finer grid spacing and consequently more memory. However, the table grid method stays computationally more efficient than the real-time carrier generation. The amount of integer operations to perform the Doppler removal operation in a complex IF architecture is given in Table 3.8.

Carrier quantization	Memory requirement
1 bit	134 kB
2 bit	268 kB
3 bit	402 kB

TABLE 3.7: Memory requirement for LUT method I

	Additions	Multiplications
Carrier generation	0	0
Carrier mixing	$2 \cdot N_{ch} \cdot F_s \cdot T_{int}$	$4 \cdot N_{ch} \cdot F_s \cdot T_{int}$
Rotation	$2 \cdot N_{ch} \cdot N_{cor}$	$4 \cdot N_{ch} \cdot N_{cor}$
Total	$2 \cdot N_{ch} \cdot (F_s \cdot T_{int} + N_{cor})$	$4 \cdot N_{ch} \cdot (F_s \cdot T_{int} + N_{cor})$

TABLE 3.8: Number of required operations for LUT method I

This method was published in [32], but the patent was withdrawn. The pre-generation of the carrier is very popular for the implementation of a software receiver [33], [34], [35] and it allows avoiding the power-hungry real-time computation of the carrier replicas.

The lookup table method is a computationally very interesting alternative to the real-time generation of the carrier. However, it requires a non negligible amount of memory that increases even more with the integration time: a longer coherent integration time requires a finer grid to keep the SNR loss to a reasonable level.

Lookup method II

Based on the same principle as [30], the following method was proposed and patented by Normark in [36]. He pre-computes a set of carrier frequency candidates to be stored in a memory with a short access time. The grid spacing is selected such as to minimize the loss due to the Doppler frequency mismatch (a 10 Hz resolution is typically chosen). Furthermore, to provide a phase alignment capability of the generated carriers, a set of initial phases is also provided for each possible Doppler frequency, as illustrated in Figure 3.7.

Contrarily to the *lookup method I* and thanks to the phase alignment capability, the number of sampling points must not obligatory correspond to an entire acquisition period. Therefore, the length of the frequency candidate vector can be chosen with respect to the available memory and becomes quasi independent of the sampling frequency.

Assuming a Doppler range of ± 10 kHz, 1'000 frequency bins with a resolution of 10 Hz, 8 phase bins

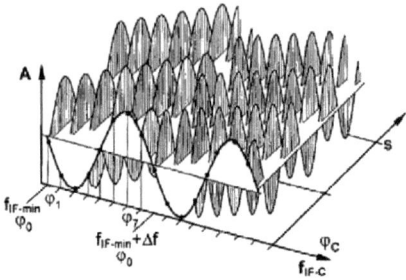

FIGURE 3.7: Set of carrier frequency candidates with initial phases [36]

with a resolution of $\pi/4$, and a vector length of 512 samples, the memory requirements are given in Table 3.9. As the memory already holds initial phase offsets with a resolution of $\pi/4$, the quadrature component of the pre-generated carrier can directly be read from the table and does not need to be stored separately.

Carrier quantization	Memory requirement
1 bit	512 kB
2 bit	1 MB
3 bit	2 MB

TABLE 3.9: Memory requirement for LUT method II

This method is an alternative implementation of the *lookup method I*. The finer frequency grid compensates for the Doppler frequency mismatch at the cost of increasing memory requirements. The set of different initial phase offsets offers alignment capabilities. However, the side effect of the phase discontinuities between two successive data blocks was not discussed in [36].

Advanced lookup method

The following method was introduced by Petovello in [37] (patent withdrawn). The principle is to perform the Doppler removal concurrently for all received satellite signals. The algorithm is implemented as a lookup table containing one single frequency and the carrier removal process is performed for all channels with the same unique frequency. Considering a Doppler range of ±10 kHz and a typically integration time of 1 ms, the frequency error results in unacceptable losses of the SNR. To overcome this problem, the integration interval T_{int} is split into M sub-intervals, as described in Equation 3.6.

$$\tilde{I}_{kl}(t) = \sum_{n=1}^{M} \tilde{I}_{kl}^n(t)$$
$$\tilde{Q}_{kl}(t) = \sum_{n=1}^{M} \tilde{Q}_{kl}^n(t)$$
(3.6)

For each sub-interval, the partial accumulation is computed and the result is rotated proportionally to the frequency mismatch. The algorithm can be applied recursively and with the adequate selection of M, the total attenuation factor can be limited to a reasonable value. A detailed mathematical development is given in [37]. Compared to the *lookup method*, the author claims an improvement of up to 30% with respect to the total complextiy for both Doppler removal and correlation stages.

Regarding the computational complexity, the Doppler removal stage remains unchanged with the difference that the operation is only performed once for all satellites while the rotation needs be executed for each of the M sub-intervals. The total amount of needed operations is given in Table 3.10.

	Additions	Multiplications
Carrier generation	0	0
Carrier mixing	$2 \cdot F_s \cdot T_{int}$	$4 \cdot F_s \cdot T_{int}$
Rotation	$2 \cdot N_{ch} \cdot N_{cor} \cdot M$	$4 \cdot N_{ch} \cdot N_{cor} \cdot M$
Total	$2 \cdot (F_s \cdot T_{int} + N_{ch} \cdot N_{cor} \cdot M)$	$4 \cdot (F_s \cdot T_{int} + N_{ch} \cdot N_{cor} \cdot M)$

TABLE 3.10: Number of required operations for advanced LUT method

This algorithm remains difficult to implement in a software receiver. First, the number of samples in one of more full C/A code varies with time and consequently, it is not possible to hardcode the number of values N_s. Second, the data samples are organized in word which will generally not be aligned to the boundaries of $M \cdot N_s$ samples. Third, the number of samples may not be divided by an arbitrary number of sub-intervals. The algorithm is therefore slightly modified in order to accommodate this and a mean power loss of 1 dB is admitted.

Although the theoretical concept is promising, the practical issues may limit the implementation of this method. The efficiency of the algorithm still needs to be evaluated in more detail, especially regarding the losses of sensitivity introduced by the multiple rotations. Furthermore, the practical implementation is not straightforward and requires some modifications that introduce further sensitivity losses.

3.7.3 Code generation and mixing algorithms

The local generation of the PRN code sequences is necessary to be able to perform the code removal.

Real-time generation and mixing with integer arithmetic

The pseudorandom noise (PRN) codes transmitted by the GPS satellites are deterministic sequences with noise-like properties. Each C/A code can be generated in real-time using a tapped linear feedback shift register. But in order to save processing power, it is preferable for software receivers to compute the 32 codes offline and to store them in memory. By the means of a NCO, the code chip sequence is oversampled at the sampling frequency F_s and then generated at the desired frequency of $F_{code} + F_{dop}/(F_s/F_{code})$.

In the same manner as for the carrier generation, the number of needed operations for a complex IF architecture is given in Table 3.11. The number of logical operations corresponds to $3 \cdot N_{ch} \cdot F_s \cdot T_{int}$ and is not included in the table for a clearer view.

	Additions	Multiplications
Code generation	$N_{ch} \cdot F_s \cdot T_{int}$	0
Code mixing	0	$2 \cdot N_{ch} \cdot N_{cor} \cdot F_s \cdot T_{int}$
Code accumulation	$2 \cdot N_{ch} \cdot N_{cor} \cdot F_s \cdot T_{int}$	0
Total	$N_{ch} \cdot F_s \cdot T_{int} \cdot (1 + 2 \cdot N_{cor})$	$2 \cdot N_{ch} \cdot N_{cor} \cdot F_s \cdot T_{int}$

TABLE 3.11: Number of required operations for code generation and mixing with integer arithmetic

The real-time generation of the local code replica is computationally expensive and is consequently seldom used this way for the implementation of software receivers. However, SIMD operations (see Section 3.8.1) can be advantageously used to parallelize the computation and speed up the code generation and mixing.

Lookup method with oversampled representation of the code

The following method was first proposed by Ledvina in [30]. The 32 PRN codes are generated offline, sampled at F_s, and stored in their oversampled representation in memory. The term oversampled comes from the fact the the sampling frequency F_s is higher than the nominal chip rate of 1.023 MHz, as shown in Figure 3.8.

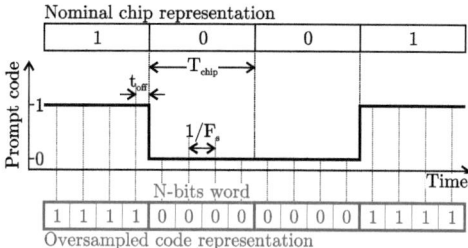

FIGURE 3.8: Oversampled code representation

The generated table also includes a selection of m different code sampling offsets t_{off} allowing to align the code precisely enough, as given in Equation 3.7.

$$t_{off} = \frac{k}{F_s \cdot m} \quad \text{for } 1 \leq k \leq m. \tag{3.7}$$

where
- m number of different sampling phases;
- k integer value;

The offset grid spacing is chosen to be large enough to guarantee sufficient timing resolution for the tracking while keeping the table at a reasonable size. For GPS applications, m is typically chosen such that the precision PR of the code alignment is in the range of a few meters, as given in Equation 3.8.

$$PR = \frac{c}{F_s \cdot m} \text{ [m]} \quad \text{with } c: \text{ speed of the light [m/s]} \tag{3.8}$$

Furthermore, as the code phase of the incoming signal is completely random, the beginning of the first code chip is most probably not synchronized with the beginning of a word and may occur anywhere within it. This can be solved either by storing all the n_s possible code phases in the memory or by shifting the code appropriately during the tracking, as illustrated in Figure 3.9. While the first solution increases the memory requirements by a factor n_s, the second requires further data processing (from 1 to n_s logical shift operations for each single word).

Regarding the Doppler frequency compensation, all the PRN codes in the table are assumed to have zero Doppler shift. The code phase errors due to this hypothesis are eliminated by choosing a replica code from the table whose midpoint occurs at the desired midpoint time. The only other effect of the zero Doppler frequency shift assumption is a small correlation power loss, which is not more than 0.014 dB (for an integration time of 1 ms) if the magnitude of the true Doppler frequency shift is less than 10 kHz [32].

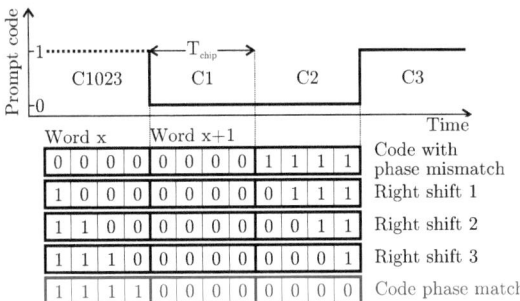

FIGURE 3.9: Code alignment (shifting by samples)

With a sampling frequency of $F_s = 8$ MHz, the memory requirements for this method to save the 32 PRN codes is given in Table 3.12.

Alignment method	Number of code offsets	Position resolution	Memory requirement
Logical shift	12	3.12 m	1 MB
Full phases storage	12	3.12 m	32 MB

TABLE 3.12: Memory requirement for saving the oversampled version of all 32 PRN codes

The number of needed operations depends on the selected code alignment method. By storing all possible code phases, the code generation is reduced to simple memory accesses. On the other hand, the logical shift method requires further logical operations for each word. The number of needed operations is given in Table 3.13.

Alignment method	Memory accesses	Logical operations
Logical shift	$N_{ch} \cdot N_{cor} \cdot F_s \cdot T_{int}/n_s$	0
Full phases storage	$N_{ch} \cdot N_{cor} \cdot F_s \cdot T_{int}/n_s$	$N_{ch} \cdot N_{cor} \cdot F_s \cdot T_{int}/2$

TABLE 3.13: Number of required operations for code generation with LUT

This method was published in [32], but the patent was withdrawn. The pre-generation and storage of the code is very popular for the implementation of a software receiver [35], [38], [39], [40], and [41].

This technique overcomes the high computational load inherent to the real-time generation of the code. On the other hand, it requires an important amount of additional memory to save all the code phases.

Real-time generation with oversampled version of the code

The following algorithm was introduced by Psiaki in [40] and by Ledvina in [42]. The real-time generation of the GPS L2 codes is required as it becomes impractical to store them entirely in an oversampled form (as the CM and CL codes have a length of 10'230 and 767'250, respectively). The principle consists in using a lookup table for translating any non-oversampled L chip sequence and its phase offset into its oversampled representation. The L value corresponds to the maximum number of chips spanning an entire data word of n_s bits, as depicted in Figure 3.10.

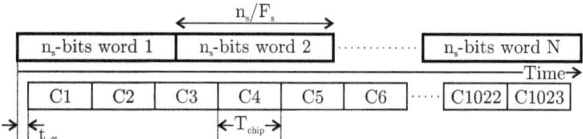

FIGURE 3.10: Relation between code chips and n_s bits data words

For each of the 2^L possible chip sequences, the table also includes a selection of sampling offsets to perform the code alignment with a time resolution of F_s/m. The offset t_{off} can then be chosen such as the Equation 3.9 is fulfilled.

$$-\frac{t_{E-L}}{2} < t_{off} < T_{chip} - \frac{t_{E-L}}{2} \qquad t_{off} = \frac{k}{F_s \cdot m} \qquad (3.9)$$

with

t_{E-L} \qquad separation between the early and late code replicas;

$1/(F_s \cdot m)$ \qquad table offset grid space.

Equation 3.9 keeps the offset range within a chip and guarantees that the start of the first late chip occurs not later than the first sample and that the end of the first late chip occurs not earlier than the first sample. This relation determines a set of $k_{tot} = k_{max} - k_{min} + 1$ possible values of k and allows the computation of L as given in Equation 3.10.

$$L = \left\lfloor \left(\frac{n_s - 1}{F_s \cdot T_{chip}} - \frac{k_{min}}{m \cdot F_s \cdot T_{chip}} + \frac{t_{E-L}}{2 \cdot T_{chip}} \right) \right\rfloor + 2 \qquad (3.10)$$

The size of the table can now be determined with the parameters k_{min}, k_{max}, and L. The table contains a selection of k_{tot} sampling offsets associated to the 2^L possible chip sequences, providing a total of $k_{tot} \cdot 2^L$ entries for each version of the code (P, E, and L). The table is organized as a counter with the first 2^L entries tabulating the 2^L possible chip sequences associated to the offset $t_{off\ min}$ and so on, as shown in Table 3.14.

	Table entries	Table outputs
Code phase offset	Non-oversampled L chip sequences	Oversampled code
$k_{min} \cdot F_s/m$	00 $\cdots$ 00	Out(0)
$k_{min} \cdot F_s/m$	00 $\cdots$ 01	Out(1)
$\vdots$	$\vdots$	$\vdots$
$k_{min} \cdot F_s/m$	11 $\cdots$ 11	Out($2^L - 1$)
$\vdots$	$\vdots$	$\vdots$
$k_{max} \cdot F_s/m$	00 $\cdots$ 00	Out($k_{tot} \cdot 2^L - k_{max}$)
$k_{max} \cdot F_s/m$	00 $\cdots$ 01	Out($k_{tot} \cdot 2^L - k_{max}+1$)
$\vdots$	$\vdots$	$\vdots$
$k_{max} \cdot F_s/m$	11 $\cdots$ 11	Out($k_{tot} \cdot 2^L - 1$)

TABLE 3.14: Real-time code generation with LUT method

Thus, given any phase offset and chip sequence C_i, the corresponding table index can be computed as given in Equation 3.11.

$$\text{Index}(k) = (k - k_{min}) \cdot 2^L + \sum_{i=1}^{L} C_i \cdot 2^{L-i} \qquad (3.11)$$

From Equation 3.11, it is also possible to compute offline all the $k_{tot} \cdot 2^L$ table outputs. The challenge consists in efficiently determining the code phase offsets in order to compute the table index corresponding to each data word; this can be performed with a set of recursive iterations as described in [42].

With a sampling frequency of $F_s = 8$ MHz, the memory requirements for this method to save the tables for the P, E, and L versions of the code are given in Table 3.15.

Parameter	Value
Number of subdivisions per sample period	12 m
Number of phase offsets t_{off}	96
Word length n_s	32
Number of chips per word L	6
Number of table entries	8192
Memory requirement	96 kB

TABLE 3.15: Memory requirement for real-time code generation with LUT method

In addition to the lookup table, the algorithm still needs to compute each word index. The author estimates the computational load to 36 integer operations per word and per version of the code.

This method was published in [32], but the patent was withdrawn. As the algorithm was introduced for the development of a dual (C/A and P code) receiver where the saving of long codes is not possible, it provided no real advantage for the C/A compared to the pre-computing method.

3.7.4 Serial search architecture

The serial search architecture performs a two-dimensional search: a code phase sweep over all 1023 different code phases and a frequency sweep over all possible carrier frequencies (affected by a Doppler offset), as shown in Figure 3.11. The serial search architecture is depicted in Figure 2.5.

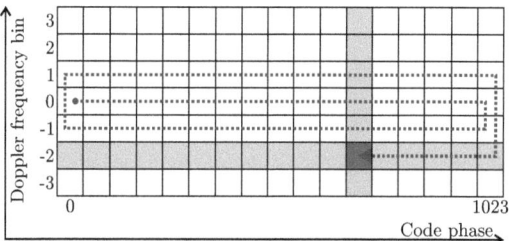

FIGURE 3.11: Serial search architecture search space

The equations to calculate the amount of integer operations is given in Table 3.16.

	Additions	Multiplications
Carrier mixing	0	$2 \cdot N_{ch} \cdot N_{bin} \cdot N_{\phi} \cdot F_s \cdot T_{int}$
Code mixing	0	$2 \cdot N_{ch} \cdot N_{bin} \cdot N_{\phi} \cdot F_s \cdot T_{int}$
Integration	$2 \cdot N_{ch} \cdot N_{bin} \cdot N_{\phi} \cdot F_s \cdot T_{int}$	0
Squaring	$1 \cdot N_{ch} \cdot N_{bin} \cdot N_{\phi} \cdot F_s \cdot T_{int}$	$2 \cdot N_{ch} \cdot N_{bin} \cdot N_{\phi} \cdot F_s \cdot T_{int}$
Total	$3 \cdot N_{ch} \cdot N_{bin} \cdot N_{\phi} \cdot F_s \cdot T_{int}$	$6 \cdot N_{ch} \cdot N_{bin} \cdot N_{\phi} \cdot F_s \cdot T_{int}$

TABLE 3.16: Number of required operations for serial search architecture

Due to the high requirements in term of the needed amount of operations, the implementation of the serial search architecture in a software receiver solution was never mentioned.

3.7.5 Parallel code search architecture

The parallel code search architecture looks for a correlation peak in the frequency domain by testing all code phases in parallel for a given Doppler frequency. This is done by performing a circular cross correlation of the input signal and the locally generated PRN code in the frequency domain. After the correlation, an inverse FFT is performed to determine if a correlation peak is present. If not, the operation is repeated with the next Doppler frequency bin by shifting the FFT components of the locally generated PRN code. The parallel code search architecture scheme is depicted in Figure 2.7. As the Fourier transform of the locally generated PRN codes can be pre-computed and stored offline, each frequency bin search consists in performing one Fourier transform and one inverse Fourier transform.

If the number of samples in a period of the PRN code is not a power of 2 (what is normally the case), the input signal must be carefully arranged with zero padding [43].

Under the assumption that N is a power of 2, the complexity (either multiplications or additions) of a N point FFT can be estimated as: $5 \cdot N \cdot \log_2(N)$ (radix-2 algorithm (Cooley and Tukey)) [44].

Furthermore, the uncertainty of a possible data bit transition for integration times longer than 1 ms has to be tackled. It might be necessary to run the acquisition algorithm twice for each acquisition step to ensure that one set of data will be free of a data bit transition (*alternate half bit method*). As the data bit transition will not be considered in the following estimations, the alternate half bit method will not be explained further.

The frequency shift into baseband (i.e., multiplication with a complex plane wave in the time-domain) can also be performed in the frequency-domain as well. The multiplication will be translated into a convolution with a single Dirac function centered at the $F_{if} + F_{dop}$ frequency. This corresponds to perform a simple cyclic shift of the FFT frequency components (as the spectrum of a FFT is periodic). However, as the Dirac function can only be positioned with a limited frequency precision, a quantization is therefore introduced on the demodulation frequency. The resolution Δf of the position can be defined as

$$\Delta f = \frac{F_s}{N} \qquad (3.12)$$

This means that a smaller FFT size N will result in a coarser frequency resolution.

Whenever a FFT is used with a rectangular time window, the frequency response is a sinc function $(\sin(x)/x)$. There is always the possibility that an input frequency will fall exactly between two frequency bins, where $x = \pi/2$. When this situation occurs, the amplitude of the signal will drop corresponding to Equation 3.13 [11].

$$20 \cdot \log\left(\frac{\sin(x)}{x}\right)\bigg|_{x=\pi/2} = 20 \cdot \log(0.6366) = -3.92 \text{ dB} \qquad (3.13)$$

The loss L associated to the frequency mismatch δf coherently integrated over T_{int} is given in Equation 3.14.

$$L_{dB} = 20 \cdot \log_{10}\left[\frac{\sin(\pi \cdot \delta f \cdot T_{int})}{\pi \cdot \delta f \cdot T_{int}}\right] \qquad (3.14)$$

Of course, the amplitude loss caused by the Doppler frequency mismatch must be added to this value.

In the absolute worst case, the maximum frequency error is half a Doppler search bin plus half a DFT frequency bin.

Assuming that the baseband mixing is performed in the frequency domain by a cyclic shift and that the FFT codes are pre-computed, the equations to calculate the amount of integer operations for the parallel code search algorithm are given in Table 3.17.

	Additions	Multiplications
FFT	$5 \cdot N_{bin} \cdot N \cdot \log_2(N)$	$5 \cdot N_{bin} \cdot N \cdot \log_2(N)$
IFFT	$5 \cdot N_{ch} \cdot N_{bin} \cdot N \cdot \log_2(N)$	$5 \cdot N_{ch} \cdot N_{bin} \cdot N \cdot \log_2(N)$
Mixing	$2 \cdot N_{ch} \cdot N_{bin} \cdot N$	$4 \cdot N_{ch} \cdot N_{bin} \cdot N$
Squaring	$1 \cdot N_{ch} \cdot N_{bin} \cdot N$	$2 \cdot N_{ch} \cdot N_{bin} \cdot N$
Total	$5 \cdot N_{bin} \cdot N \cdot \log_2(N) \cdot (N_{ch}+1) + 3 \cdot N_{ch} \cdot N_{bin} \cdot N$	$5 \cdot N_{bin} \cdot N \cdot \log_2(N) \cdot (N_{ch}+1) + 6 \cdot N_{ch} \cdot N_{bin} \cdot N$

TABLE 3.17: Number of required operations for parallel code search architecture

The parallel code search architecture is widely used in the implementation of software receivers and referenced in may papers such as [8], [23], [41], [45], [46], [47], [48], [49], [50], and [51].

3.7.6 Parallel frequency search architecture

The parallel frequency search architecture looks for a correlation peak in the frequency domain by testing all Doppler bins at once for a given code phase. The baseband signal is multiplied with the locally generated PRN code in order to form P consecutive partial correlations with a pre-detection time T_{coh} which is P times smaller than the integration time T_{int}. The P results are then regrouped in a vector on which a N-point FFT is computed where $N \geq P$ (if $N > P$, zero padding has to be applied). Assuming P large enough, all Doppler bins are searched in parallel for a given code phase. If no correlation peak is detected, the operation is repeated with the next code phase. The parallel frequency search architecture scheme is depicted in Figure 2.6.

The architecture is characterized by the parameters given in Table 3.18.

Parameter	Description
$P \cdot T_{coh}$	Total coherent integration time T_{int}
$1 / T_{coh}$	New sampling rate and search bandwidth of the FFT
$1 / P \cdot T_{coh}$	Frequency bin resolution (bin width)

TABLE 3.18: Parameters for parallel frequency architecture

The FFT should have frequency components covering the range of ± 10 kHz in order to find the Doppler shift in one step. Therefore, the signal should be down-sampled at a frequency of 20 kHz (i.e., using a pre-detection time of $T_{coh} = 50~\mu s$). Typical FFT configurations are given in Table 3.19.

	P	T_{coh} [μs]	T_{int} [ms]	Search bandwidth [kHz]	Frequency resolution [Hz]
Reduced range	4	250	1	± 2	1000
Full range, low sensitivity	20 (32)	50	1	± 10	625
Full range, average sensitivity	32	50	1.6	± 10	625
Full range, high sensitivity	256	50	12.8	± 10	78.125

TABLE 3.19: Typical configurations of parallel frequency search architecture

The equations to calculate the amount of integer operations is given in Table 3.20.

	Additions	**Multiplications**
Baseband mixing	0	$2 \cdot N_{ch} \cdot N_\phi \cdot F_s \cdot T_{int}$
Code mixing	0	$2 \cdot N_{ch} \cdot N_\phi \cdot F_s \cdot T_{int}$
Integration	$2 \cdot N_{ch} \cdot N_\phi \cdot F_s \cdot T_{int}$	0
FFT	$N_{ch} \cdot N_\phi \cdot 5 \cdot N \cdot \log_2(N)$	$N_{ch} \cdot N_\phi \cdot 5 \cdot N \cdot \log_2(N)$
Total	$N_{ch} \cdot N_\phi \cdot (5 \cdot N \cdot \log_2(N) + 2 \cdot F_s \cdot T_{int})$	$N_{ch} \cdot N_\phi \cdot (5 \cdot N \cdot \log_2(N) + 4 \cdot F_s \cdot T_{int})$

TABLE 3.20: Number of required operations for parallel frequency search architecture

The parallel frequency search architecture is rarely found in the descriptions of the implementation of a software receiver. Most of the time, it is combined with the parallel code search architecture in order to improve the Doppler estimation as mentioned in [48].

3.7.7 Comparison of the different acquisition methods

To compare the the results of the different acquisition algorithms presented above, the following assumptions are made:

The complexity for the different acquisition algorithms are given in Table 3.21 for one millisecond integration time and in Table 3.22 for 10 milliseconds integration time.

Chapter 3 Software receivers 63

Number of channels: 12
Sampling frequency: 4 MHz (step size of 1/4 chip)
Doppler search space: ± 5 kHz
Integer additions and multiplications

	additions	multiplications
Serial search	$8.45 \cdot 10^9$	$1.69 \cdot 10^{10}$
Parallel code search	$1.67 \cdot 10^8$	$1.74 \cdot 10^8$
Parallel frequency search	$1.08 \cdot 10^8$	$2.06 \cdot 10^8$

TABLE 3.21: Comparison of numerical values of required operations for different acquisition architectures (1 ms integration time)

	additions	multiplications
Serial search	$8.45 \cdot 10^{11}$	$1.69 \cdot 10^{12}$
Parallel code search	$3.53 \cdot 10^9$	$3.64 \cdot 10^9$
Parallel frequency search	$9.91 \cdot 10^8$	$1.97 \cdot 10^9$

TABLE 3.22: Comparison of numerical values of required operations for different acquisition architectures (10 ms integration time)

The serial search approach is generally preferred in hardware receivers where the high parallelism capabilities are exploited by implementing many code replicas (also called correlators) concurrently. A hardware implementation of the parallel code search method would be penalized by the relatively low speed of the clock (typically a few ten of MHz) and the prohibitive silicon area requested for the large FFTs. However, the post-correlation FFT requires only a small silicon area overhead, as only short FFTs are required (typically 256 points or smaller), and no significant signal storage is necessary. In the case of a software receiver, all the operations are executed sequentially and performing a serial search would result in a huge amount of computations. Consequently, FFT based acquisition methods are preferred for their lowest processing load; large FFTs can easily be implemented and can take advantage of the high CPU processing speed.

3.7.8 FFT performance values

Pany made in his latest book an overview of the reported FFT performance values of current CPUs [13].

To be able to sucessfully implement correlators working in the frequency domain, an efficient FFT method is required. The computational load (i.e., the number of needed operations (either additions or multiplications)) of a complex FFT of size N is approximately given by Equation 3.15 [44].

$$\#Ops_{FFT} = 5 \cdot N \cdot \log_2(N) \tag{3.15}$$

More specifically, there are $N\cdot\log_2(N)$ complex additions required, as well as $(N/2)\cdot\log_2(N)-3/2\cdot N+2$ complex multiplications [52].

A FFT performance test for a complex FFT with 32 bit floating-points values and with 16 bit fixed-point values were made by T. Pany and gave typical performance values of around 2.4 – 2.8 GOPS (giga operations per second; either floating-point or fixed) for the test system described in Table 3.23. The well-performing FFT library from Intel was used [53]. A $N = 8'192$ point FFT can thus be calculated within around 200 μs. The performance depends on the FFT size and a relative performance reduction can be seen if the FFT exceeds a value of $2^{16} = 65'536$ points. The limiting factor may be the integrated L2-cache of the processor (2 MB). Reported FFT performance values for other computer systems can be found in the publication by Frigo and Johnson [54]. A 3.0 GHz Intel Core 2 Duo, for example, can achieve a performance of 12.5 GOPS for a complex 32 bit floating-point FFT with $N = 65'536$, being five times faster than the test system of Table 3.23. Thus, a FFT of $N = 8'192$ points can be calculated in roughly 40 μs.

Parameter	Test system
CPU	Intel Pentium M (780)
CPU clock speed	2.26 GHz
CPU architecture	IA-32

TABLE 3.23: FFT test system configuration

Then, Pany extended the discussion by comparing the frequency-domain and the time-domain correlations in terms of operations (i.e., multiplications and additions) required to obtain the same result. This allows to determine the so called *number of equivalent correlators*. Frequency-domain techniques gain performance with an increasing number of correlators. Their use in signal acquisition is a common practice in GNSS software receivers. He made the following example using a conventional frequency technique, namely zero padding.

If M correlation values of two real-valued signals of length N shall be calculated in the time-domain, then for each sum a number of N multiplications and N additions have to be performed, resulting in

$$\#Ops_{time} = 2 \cdot M \cdot N \tag{3.16}$$

operations (see Figure 3.12). Here, the assumption was made that the first signal is of a fixed length and it is not periodic. For each correlation value, the second signal is taken from a vector of size

$M + N − 1$, each time shifted by one sample. This is the case in a GNSS receiver when correlating the incoming signal with a locally generated PRN code sequence. Further, it was assumed that $M \leq N$ and that $M + N − 1$ is an integer power of two.

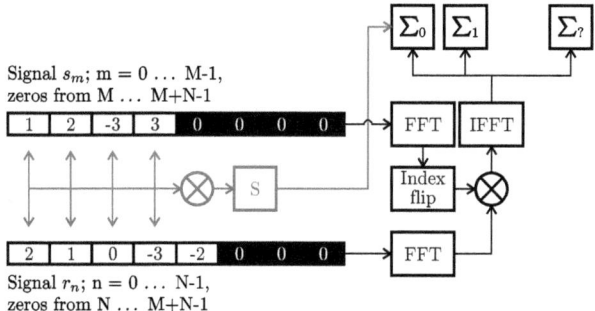

FIGURE 3.12: Frequency-domain correlator with zero padding (light-gray elements symbolize the time-domain correlation for comparison; the symbol $\sum$ denotes an accumulator)

To achieve the same result with frequency-domain techniques, the following steps have to be performed (see Figure 3.12):

1. Zero pad the first signal to a length of $M + N − 1$;
2. Forward FFT the real-valued zero-padded first signal;
3. Forward FFT the real-valued second signal;
4. Frequency-domain multiplication;
5. Inverse FFT of the frequency-domain product.

A real-valued FFT needs half the operations as a complex-valued FFT, thus (2), (3), and (5) need all together

$$\#Ops_{FFT,1} = 10 \cdot (M + N − 1) \cdot \log_2(M + N − 1) \tag{3.17}$$

operations. The spectral multiplication needs

$$\#Ops_{FFT,2} = M + N − 1 \tag{3.18}$$

operations, yielding to an overall number of operations of

$$\begin{aligned}\#Ops_{FFT} &= \#Ops_{FFT,1} + \#Ops_{FFT,2} \\ &= (M + N − 1) \cdot (1 + 10 \cdot \log_2(M + N)).\end{aligned} \tag{3.19}$$

The last expression shows that for large N, the number of operations increases with $N \cdot \log_2(N)$, whereas in the time domain the increase is quadratic (assuming M proportional to N).

The ratio between the time span represented by N samples divided by the CPU time of the FFT implementation to complete the correlation defines the correlation efficiency of an FFT-based correlator algorithm. The effective number of correlators $\#Corr$ is defined as the product of correlation efficiency multiplied with N. The effective number of correlators corresponds to the number of hardware correlators that would give the same result within the same time. It is important to note that hardware correlators are understood to work intrinsically in real time (and not faster). The definition of $\#Corr$ assumes that all available correlation values from the frequency-domain correlation are exploited (i.e., $M = N$ is assumed for the most effective use of the frequency-domain correlation). The effective number of correlator is therefore

$$\begin{aligned}\#Corr &= N \cdot \frac{T_{coh}}{T_{CPU}} = \frac{N \cdot T_{coh} \cdot f_{OPS}}{2 \cdot N \cdot (10 \cdot \log_2(2 \cdot N) + 6)} \\ &= \frac{T_{coh} \cdot f_{OPS}}{2 \cdot (10 \cdot \log_2(N) + 16)}.\end{aligned} \quad (3.20)$$

Here, T_{coh} denotes the length of the signal in seconds (usually the coherent integration time given as the number of samples N divided by the sample rate). T_{CPU} is the time of the CPU to perform the computation in seconds, f_{OPS} is the number of FFT operations the CPU can perform within one second. The longer the coherent integration time T_{coh}, the more efficient the FFT.

If an example of a 20-ms signal, a FFT size of $N = 2^{15}$ points, and a FFT performance of $f_{OPS} = 2.5$ GOPS is taken, the effective number of correlators gives 150'602. If a Doppler frequency preprocessing can be applied, the number of effective correlators further increases (e.g., by a factor of approximately 10-20 for the case of the GPS C/A-code).

3.8 Alternate processing methods

A major problem with the software architecture are the important computing resources required for the baseband processing. As a straightforward transposition of traditional hardware based architectures into software would lead to an amount of operations which is not suitable for today's fastest computers, three main alternate strategies can be found in the literature:

1. Using SIMD operations;
2. Using vector or bitwise processing;
3. Using additional resources (like Graphics Processing Unit (GPU)).

The first one relies on the utilization of SIMD operations that provide the capability of processing vectors of data. Since they operate on multiple integer values at the same time (contrary to Single

Chapter 3 Software receivers 67

Instruction Single Data (SISD)), SIMD could result in significant gains in execution speed for repetitive tasks such as baseband processing. However, SIMD operations are tied to specific processors and therefore severely limit the portability of the code.

The second alternative consists in the parallel bitwise operations (sometimes also referred as vector processing in the literature), which exploit the native bitwise representation of the signal samples. The data bits are stored in separate vectors, one sign and one or several magnitude vectors on which bitwise parallel operations can be performed. The objective is to take advantage of the universality, high parallelism, and speed of the bitwise operations for which a single integer operation is translated into a few simple parallel logical relations. While SIMD operations use advanced and specific optimization schemes, the latter methodology exploits universal CPU instructions set.

The third method is the most recent one and uses additional high performance resources (like a GPU) that are available in a standard personal computer for the computation of the heavy calculations.

These three methods will be shortly discussed in the following sections.

3.8.1 Single Instruction Multiple Data (SIMD)

In 1995, Intel introduced the first instance of SIMD under the name of Multi Media Extension (MMX) [55]. The SIMD are mathematical instructions that operate on vectors of data and perform integer arithmetic on eight 8 bit, four 16 bit, or two 32 bit integers packed into a MMX register (see Figure 3.13).

On average, the SIMD operations take more clock cycles to execute than a traditional x86 operation. Anyhow, since they operate on multiple integers at the same time, MMX code can result in significant gains in execution speed for appropriately structured algorithms. Later SIMD extensions (SSE, SSE2, SSE3, and SSE4) added eight 128 bit registers to the x86 instruction set. Additionally, SSE operations include SIMD floating point operations and expand the type of integer operations available to the programmer.

SIMD operations are well fitted to parallelize the operations of the baseband processing (BBP) stage. In particular, they can be used to allow the PRN code mixing and the accumulation to be performed concurrently for all the code replicas. With the help of further optimizations such as instruction pipelining, more than 600% performance improvement with the SIMD operations compared to the standard integer operations can be observed [29]. For this reason, most of the software receivers with real-time processing capabilities use SIMD operations [29], [38], [33], [34].

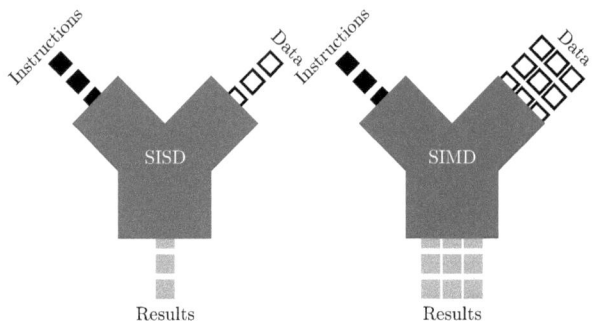

FIGURE 3.13: SISD versus SIMD

3.8.2 Bitwise (vector processing) operations

Bitwise operation (or vector processing) was first introduced in [30]. The method exploits the bit representation of the incoming signal where the data bits are stored in separate vectors on which bitwise parallel operations can be performed. Figure 3.14 shows a typical data storage scheme for vector processing.

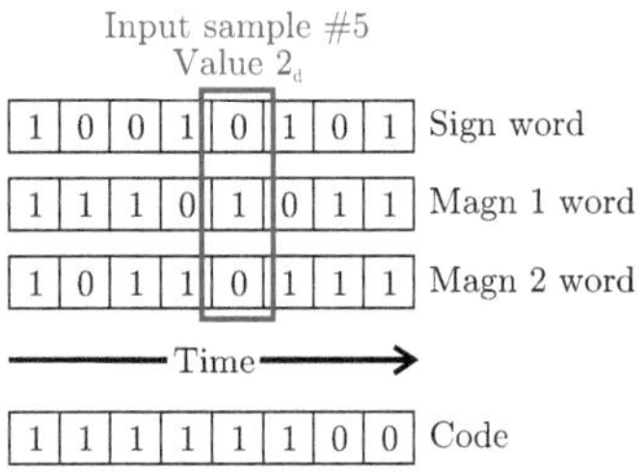

FIGURE 3.14: Bitwise data representation

The sign information is stored in the *sign word* while the remaining bit(s) representing the magnitude is (are) stored in the *magn word(s)*. The objective is to take advantage of the high parallelism and speed of the bitwise operations for which a single integer addition or multiplication is translated into simple parallel logical operations. The carrier mixing stage is reduced to one or a few simple logical operations which can be performed concurrently on several bits. In the same way, the PRN code removal only affects the sign word.

In [32] the complete code and carrier removal process requires two operations for each code replica (Early, Prompt, and Late). The complexity can be even further reduced by more than 30% by considering one single combination of early and late code replicas (typically early-minus-late). This way, the author claims an improvement of a factor 2 for the bitwise method compared to the standard integer operations.

The inherent drawback of this approach is the lack of flexibility: the complexity of the process becomes bit-depth dependent and the signal quantification cannot be easily changed (while performing BBP with integers allows the signal structure to change significantly without code modification).

To overcome this limitation, a combination of bitwise processing and distributed arithmetic can be used. This method was described in detail in [56]. The power consuming operations are performed with bitwise operations and to be able to keep the flexibility of the calculations standard integer operations are used after the code and carrier removal. The passage between the two methods is done with the distributed arithmetic.

3.8.3 Use of Graphics Processing Unit (GPU)

Another alternate processing method is to make use of the GPU available in every standard personal computer. These devices provide low-cost massive parallel computing performance, which can be used for the implementation of a software GNSS receiver.

Driven by the requirements of the PC gaming industry, GPUs have evolved to massive parallel processing systems which entered the area of non-graphic related applications several years ago (see [57] and [58]). Although a single processing core on the GPU is much slower and provides less functionality than its counterpart on the CPU, the huge number of these small processing entries outperforms the classical processors when the application can be parallelized. [59] demonstrate that a GPU can be successfully used for radio astronomical signal processing, solving tasks which are similar to those of a GNSS receiver. Contrarily to the CPUs which directly access the PC's memory, it is necessary to transfer the relevant data from the CPU memory to the onboard-memory of the graphic card, before it can be accessed by a program running on the GPU. Thus, the data transfer between the CPU and the GPU can be a significant bottleneck.

T. Hobiger presented in a paper the solution of a real-time multi-channel software receiver running on a standard GPU [60]. Basically, only off-the-shelf components have been used and the GPU code has been compiled by the help of the NVIDIA's CUDA environment [61]. The calculation of the FFT was performed by the NVIDIA's FFT library CUFFT and other functions of the CUDA toolkit turned out to be very useful for debugging the code. The exact description of the implementation will not be given here, but can be found in the reference. He demonstrated that an implementation of a real-time software receiver is possible on a GPU, yielding similar results as obtained from a hardware receiver. Currently, only moderate sampling rates can be processed by a GPU (i.e., up to 16 Msps).

3.9 Summary

The development of software radios started back in the 1990s driven by the need of an universal solution for different frequencies and modulations. The first "real" implementation of a software receiver (as given by definition) was in 1997 and from this time on, the developments made huge progresses. Today, several solutions can be found that run in real-time with more than 10 channels on standard (or embedded) microprocessors. Nevertheless, many of the existing solutions utilize some specific instructions that are tightly coupled to a microprocessor architecture.

Some of the main challenges of an implementation of a software receiver have been presented, together with numerical values that showed the number of needed operations for the different blocks and why it is difficult to implement a GNSS receiver in software.

The chapter also presented an overview of the current implementation and the different algorithms and architectures that have been specially designed for real-time capable software receivers. This included the architecture of the baseband stage as well as the algorithms for the acquisition and tracking and the local generation of the carrier and the code.

This chapter contains the following contributions worked out and partially published during the elaboration of this thesis:

- The collection of the different definitions and types of software receivers in the domain of GNSS;
- The survey of the main challenges for software receivers;
- The extensive compilation and comparison of current implementation and algorithms for software receivers, including their complexities and the amount of needed operations;
- The proposition for using distributed arithmetics in the bitwise operations.

Chapter 4

New architecture and algorithms for a software receiver

4.1 Introduction

This chapter will describe the new architecture for the implementation of a real-time capable software receiver that was developed in the scope of this thesis. This includes the discussion about where to separate between hardware and software, i.e., which part should be implemented in hardware and which part in software. Then, the different components of the proposed architecture are explained and described, including the connection between the different elements.

4.2 Separation hardware / software

Although an ideal software receiver would only consist of an antenna and an ADC, this solution is currently not yet possible as discussed in Section 3.6. Therefore, a compromise has to be made and some operations still need to be realized in hardware. Furthermore, the computational load of the software receiver can be drastically reduced by putting some key elements into additional hardware. This section will explain the selected separation of hardware and software. The implementation of the software receiver (see Chapter 5) and the obtained results (see Chapter 6) are based on this architecture.

The conversion of the analog incoming signal to digitized samples has to be realized in hardware, as described in Section 4.3.1. In order to reduce the computational load of the software part, an additional stage was introduced called Baseband Pre-Processing (BBPP), as described in Section 4.3.2. The data samples are then arranged properly to be handed over to the host interface controller that transmits the data stream to the software receiver. All the baseband processing operations are afterwards executed in software.

The operations given in Table 4.1 are implemented on the hardware components.

Operation	Hardware component
Conversion to low-IF	RF front-end
Analog-to-digital conversion	RF front-end
Conversion to baseband	FPGA
Bandwidth reduction	FPGA
Bit alignment	FPGA
Transfer to host computer	USB 2.0 controller

TABLE 4.1: Overview of implemented operations in hardware

4.3 New architecture

This section will present the new architecture of the software receiver, as depicted in Figure 4.1. It consists of a RF front-end unit responsible for down-converting and digitizing the incoming satellite signal and an additional BBPP stage used for relaxing the computational load of the software receiver (performing a down-conversion to baseband, a bandwidth reduction, and a bit alignment). The obtained data stream is then sent to the software receiver who perfoms the baseband processing and computes the navigation solution. The different blocks will be explained in this chapter.

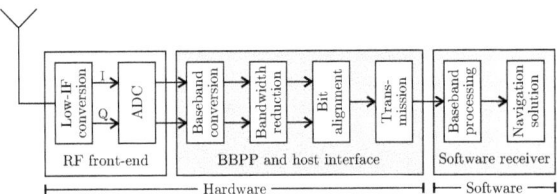

FIGURE 4.1: New software receiver architecture

4.3.1 RF front-end

The RF front-end is the first element after the antenna and converts the incoming signal into a complex signal at $F_{if} = 3.42$ MHz. The down-converted signal is then digitized with a resolution of 5 bits and a sampling frequency of 24 MHz. The obtained data stream is handed over to the BBPP stage for further pre-processing.

4.3.2 Baseband Pre-Processing

The BBPP stage is common to all channels and is implemented on a FPGA. It is responsible for following three main functions:

1. The conversion baseband;
2. The bandwidth reduction;
3. And the bit decimation.

Figure 4.2 shows the schematic block diagram of the BBPP unit.

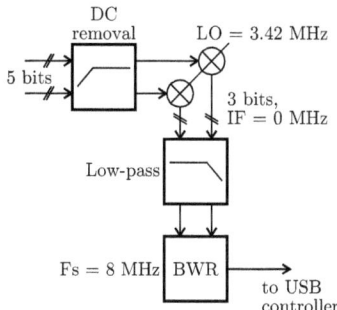

FIGURE 4.2: Baseband Pre-Processing block scheme

The digital data stream is received from the RF front-end with 5-bit quantization and a sampling frequency of 24 MHz. First, the DC component is removed by a high-pass filter and the signal is converted to baseband (multiplication with a Local Oscillator (LO) running at the intermediate frequency of 3.42 MHz). The signal is decimated to a 3 bit resolution and the sampling frequency is also lowered to 8 MHz. Finally, the incoming signal samples are re-arranged (bit alignment) and transmitted to the software receiver via the host interface (USB 2.0 interface).

The BBPP stage was provided as is and will not be further discussed in this thesis.

4.3.3 Host interface

Although a software receiver implements most of the processing algorithms in software, a hardware front-end is always needed that allows receiving the incoming signals. The RF signal is digitized and the data stream has to be transmitted over an appropriate interface to the processor that executes the signal processing algorithms. The choice of the host interface depends strongly on the sampling frequency as this parameter defines how much data must be transmitted but also on the number of quantization bits. Table 4.2 brings these parameters together and defines the needed bandwidth of the host interface.

Sampling frequency [MHz]	Quantization bits	Data rate [MBit/s]	Data rate [MBytes/s]
4	1	8	1
	2	16	2
	3	32	4
8	1	16	2
	2	32	4
	3	48	6

TABLE 4.2: Minimal data rates for host interface (complex data)

Several common and often-used interfaces are available for embedded applications. Table 4.3 gives a non exhaustive overview of the interfaces with respect to the maximum allowed transfer speed.

Interface	Data rate [MBit/s]	Data rate [MBytes/s]
Controller Area Network (CAN)	1	0.125
RS-232	1.5	0.187
Microwire	3	0.375
Inter-Integrated Circuit (I^2C)	3.4	0.425
Serial Peripheral Bus (SPI)	10	1.25
Firewire (IEEE 1394a/b)	400/800	50/100
USB 2.0	480	60

TABLE 4.3: Available host interfaces and their maximum data rates

Only Firewire and USB 2.0 can accommodate the requested data rates of Table 4.2. The USB 2.0 interface has been chosen for the prototype due to the wider availability and the existing drivers.

USB 2.0

The USB system has an asymmetric design layout consisting of a host (that is the master of the whole system), a multitude of downstream USB ports, and multiple devices connected to the tiered-star topology. A single physical USB device may consist of several logical sub-devices that are referred to as device functions. Each device may provide one or several functions, such as a webcam (video

Chapter 4 New architecture and algorithms for a software receiver   75

device function) with a built-in microphone (audio device function). It is important to notice that all communications will be initiated by the host controller and that the devices cannot communicate between each other or send an event to the host controller.

The communication with the USB devices is based on pipes (logical channels) that connect the host controller to a logical entity on the device named endpoint. While the device sends or receives data on a series of endpoints, the client software transfer data through pipes that have a set of parameters associated with them, such as the allocated bandwidth, the transfer type, the direction of the data flow, the maximum packet size, and the number of internal buffers. Each endpoint is unidirectional and can transfer data either into (IN) or out (OUT) of the host controller. Therefore, an endpoint can be seen as a source or sink of data. An USB device can have up to 32 active endpoints, 16 IN and 16 OUT, for one configuration (called configuration interfaces).

The complete USB bandwidth is divided into 1'000 frames of 1024 Bytes per second. As compared to USB 1.0, the USB 2.0 specification subdivides each frame into 8 microframes of 125 μs length to reduce the need for buffering. It also increases the number of packets from 1 to 3 per microframe for each device.

There are four different transfer types available in the USB 2.0 specification:

- *Isochronous transfer*
  Isochronous transfer occurs continuously and periodically and is mainly used for data that cannot tolerate delay and typically contain time sensitive information. This transfer method provides guaranteed access to the USB bandwidth, bounded latency, and error detection via CRC, but no retry or guarantee of delivery.
  The maximum isochronous packet size (for High-Speed devices) is 1024 Bytes and the maximum data rate is 24 MBytes/s.
- *Bulk transfer*
  Bulk transfer can be used for large burst data that cannot tolerate errors. This method uses the unallocated bandwidth on the bus after all other transactions have been allocated and should therfore only be used for time insensitive data.
  The maximum bulk packet size (for High-Speed devices) is 512 Bytes and the maximum data rate is 53.125 MBytes/s.
- *Control transfer*
  Control transfer is typically used for command and status operations and will not be discussed further.
- *Interrupt transfer*
  Interrupt transfer is typically a non-periodic communication requiring bounded latency. It is (again) important to note that the device cannot generate an event that interrupts the host controller, but the later must poll the device to see if it has an interrupt waiting for processing.

The maximum interrupt packet size (for High-Speed devices) is 1024 Bytes and the rate of polling is specified in the endpoint descriptor.

4.3.4 Hardware prototype

The prototype uses the following off-the-shelf components:

RF front-end board

A RF front-end board from u-blox AG was used for the prototype (see Figure 4.3). It consists of a PCB that was built with discrete components using a low-IF architecture. The incoming signal is sampled at 24 MHz with a resolution of 8 bit. The connection to the BBPP and host interface board is made through a high-speed RF connector from Samtec (on the bottom side of the PCB).

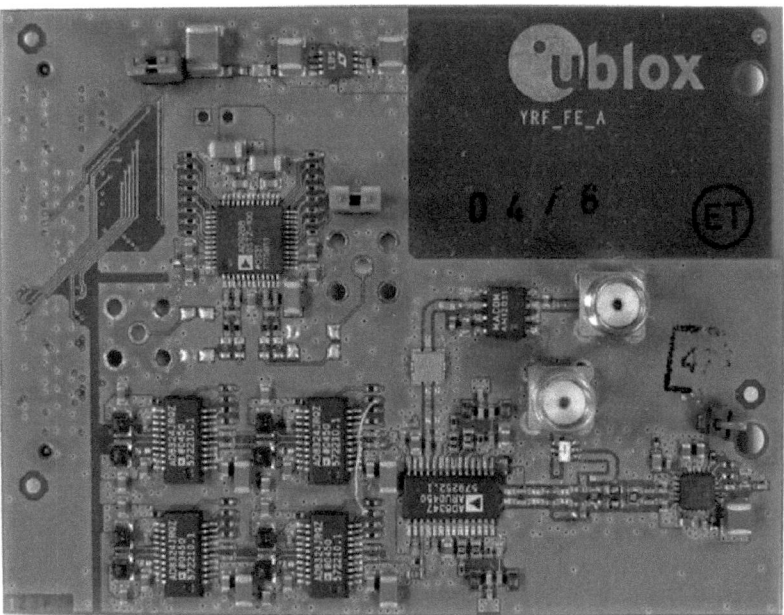

FIGURE 4.3: RF front-end board

BBPP and host interface board

A low-cost FPGA board from *FGPA-DEV* was chosen as the BBPP and host interface board (reference: *Cyclone II Board*, see Figure 4.4). It is based on an *Altera Cyclone II EP2C35* FPGA that holds the BBPP unit. It also embeds a *Cypress EZ-USB FX2LP* USB 2.0 controller (reference:

$CY7C68013A\text{-}56PVXC$) for high-speed data transfers with the host system. Furthermore, it offers enough input/output pins to connect the RF front-end board.

Details about the implementation of the BBPP stage and the configuration of the USB 2.0 controller can be found in Chapter 5.

FIGURE 4.4: FPGA-DEV Cyclone II board

4.4 New algorithms

This section will introduce the new algorithms and the final architecture for the implementation of the real-time capable software receiver. First, the general concept and the advantages of using a BBPP stage are explained. Afterwards, the new concept for generating and mixing the carrier and the code are introduced. The next part will explain the chosen architecture for the acquisition stage and the section will be concluded with the presentation of the final architecture of the software receiver. The implementation in Chapter 5 will be based on the explications given in this section. The configuration given in Table 4.4 is used as the reference for the following development and description.

Parameter	Value
Sampling frequency (F_s)	8 MHz
Integration time (T_{int})	10 ms
Signal quantization (N_q)	3 bits
Doppler frequency (F_{dop})	± 5 kHz
Quartz offset (F_{qrz})	± 45 kHz
Number of code replicas (N_{cor})	3 (Early, Prompt, and Late)
Number of channels (N_{ch})	12

TABLE 4.4: Reference receiver configuration

4.4.1 General concept of the baseband processing

In a classical approach, the conversion from the intermediate frequency into baseband is performed directly in the receiver by removing both the intermediate F_{if} and the residual Doppler F_{dop} frequencies in a single step. This requires generating the local carrier frequency $F_{if} + F_{dop}$ of a few megahertz. By introducing an intermediate BBPP stage in-between the RF board and the baseband processing stage (see Figure 4.5), the down-conversion is split into two distinct steps. The signal is first down-converted to the Doppler frequency by removing the intermediate frequency of a few megahertz. Finally, in the baseband processing (BBP) stage of the receiver, the signal is further down-converted by removing the Doppler frequency of a few kilohertz. The value of F_{if} in Figure 4.5 is only given as an example.

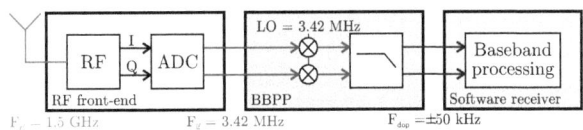

FIGURE 4.5: Frequency plan of new software receiver architecture

The implementation of the BBPP in a separate hardware relaxes the constraints on the software receiver itself as the carrier frequencies to be generated internally are in the range of a few kilohertz only instead of a few megahertz. The highest frequency to be generated is given by the satellite Doppler shift (± 5 kHz) and the quartz offset (± 45 kHz). As the latter is the same for all satellites, it can be compensated directly during the baseband conversion and consequently, only the Doppler frequency needs to be generated internally. This property is advantageously exploited for optimizing all the baseband processing operations.

4.4.2 Carrier generation and mixing

Let us consider the generation of the carrier frequency by the mean of a NCO as shown in Figure 3.6. The generated frequency F_{car} depends on the accumulator bit-width W and is proportional to the sampling clock F_s and the phase increment NCO_{inc}, as given in Equation 3.3.

The number of clock cycles needed to achieve one carrier period (i.e., when the accumulated phase overflows the accumulator capacity), is given in Equation 4.1.

$$N_{clk} = \frac{2^W}{NCO_{inc}} = \frac{F_s}{F_{car}} \qquad (4.1)$$

When generating a carrier with a frequency $|F_{car}| < 5$ kHz, the phase accumulator increases very slowly and the carrier sign and magnitude remain constant during many clock cycles. Consequently, several consecutive incoming samples can be regrouped into batches, as they are multiplied with the same carrier value. The data is said to be *batch processed*.

The average batch size $\overline{B}$ is related to the N_q bit carrier quantization, which fixes the $2^{N_{quant}+1}$ different complex carrier phases and is then given in Equation 4.2.

$$\overline{B} = \frac{N_{clk}}{2^{N_q+1}} = \frac{F_s}{2^{N_q+1} \cdot F_{car}} \qquad (4.2)$$

For a maximal Doppler frequency shift of $F_{dop} = 5$ kHz, (as given in Table 4.4), at least 100 incoming samples can be regrouped and summed up for each of the respective carrier phase k. This way, the carrier generation process is mostly reduced to the computation of the batch size $B(k)$ associated to each carrier phase, as given in Equation 4.3 (a zero initial carrier phase offset beeing assumed).

$$B(k) = \left\lfloor (k+1) \cdot \overline{B} \right\rfloor - \left\lfloor k \cdot \overline{B} \right\rfloor \qquad (4.3)$$

where $k \in [0, K-1]$

K number of carrier phases in one integration period

The batch size $B(k)$ is bound to the number of samples N_s processed per integration period, as given in Equation 4.4.

$$N_s = \sum_{k=0}^{K-1} B(k) \qquad (4.4)$$

For the configuration in Table 4.4, K is in the range of $[1; 80]$ and $B(k)$ in the range of $[8000; 100]$.

The introduction of the batch processing simplifies the carrier removal process as the order of the operation is modified. The incoming samples are first mixed with the code and pre-accumulated in $B(k)$ sample batches. This operation translates into a redistribution of the conventional baseband architecture with the carrier removal process intervening after the accumulation stage. The new architecture is shown in Figure 4.6.

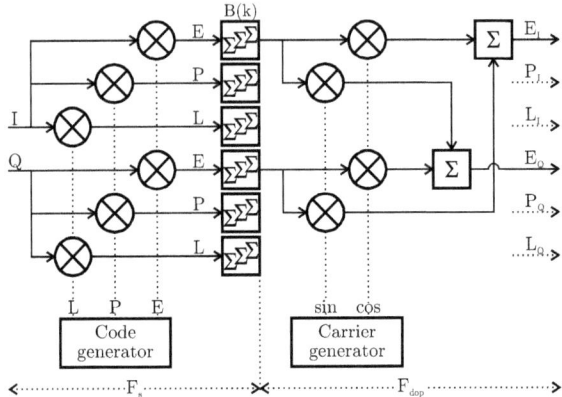

FIGURE 4.6: New baseband architecture with carrier batch processing; the multiple sum symbols indicate the partial sums over $B(k)$ samples

The incoming samples are first multiplied with the code and then accumulated into batches of length of $B(k)$ samples. This operation is performed at the sampling frequency rate F_S. The K batches are then multiplied with the respective carrier amplitude value and summed up to form the definitive correlation results E_I, P_I, L_I and E_Q, P_Q, L_Q. Consequently, the data throughput is progressively reduced starting from the sampling frequency F_S to the Doppler frequency rate F_{dop}, reducing also the amount of required operations in the carrier removal process.

As the carrier is periodic, the amount of needed operations can be reduced even more by combining every $2^{(N_q+1)\text{th}}$ batch. The sum of these batches can be multiplied once with the same corresponding amplitude value. This decreases the number of logical operations and multiplications by a factor of F_{dop} to only 2^{N_q+1} per channel.

Regarding the carrier generation and mixing, the use of batch processing leads to the amount of needed integer operations given in Table 4.5.

	Additions	Multiplications
Generation	$N_{ch} \cdot F_{dop} \cdot 2^{N_q+1}$	0
Mixing	$2 \cdot N_{cor} \cdot N_{ch} \cdot F_{dop} \cdot 2^{N_q+1}$	$4 \cdot N_{cor} \cdot N_{ch} \cdot F_{dop} \cdot 2^{N_q+1}$

TABLE 4.5: Number of required operations (per second) for carrier batch processing

The batch processing based baseband architecture offers several advantages:

Chapter 4 New architecture and algorithms for a software receiver 81

- First of all, it reduces significantly the complexity of the carrier mixing process, especially in terms of integer multiplications. This leads to a number of operations that becomes suitable for a real-time implementation on current processors.
- The use of a classical NCO allows the generation of the exact carrier replica at any desired phase and frequency. This simplifies greatly the integration of the carrier generation block in the receiver as it can be adopted easily to various tracking schemes.
- The memory requirements are reduced to the strict minimum and only 2^{N_q+1} carrier levels need to be stored.
- And last but not least, the redistribution of the complete baseband architecture opens new perspectives by expanding the concept of batch processing not only to the carrier removal process, but also to the code removal process.

4.4.3 Code generation and mixing

The code generation can also be performed the same way as the carrier generation, namely by the mean of a NCO as depicted in Figure 3.6. The average number of clock cycles needed to produce one chip period is proportional to the sampling frequency:

$$N_{clk} = \overline{P} = \frac{F_s}{F_{code}} \tag{4.5}$$

For the configuration given in Table 4.4, the average number of samples per chip is approximately 7.82 (= 8000/1023), depending on the residual Doppler frequency on the code chip rate. Consequently, several consecutive samples can be combined into batches (referenced to as *partial sums*) as they are multiplied with the same chip value. The code generation process is therefore reduced to the computation of the different partial sum sizes. The latter can be estimated recursively as given in Equation 4.6 (example with zero initial code phase offset).

$$P_L(j) = \lfloor (j+1) \cdot \overline{P} \rfloor - \lfloor j \cdot \overline{P} \rfloor \tag{4.6}$$

For example, with the nominal sampling and frequency rate, the number of samples contained in the first 12 partial sums is given in Equation 4.7.

$$P_L(0), \ldots, P_L(11) = 7\ 8\ 8\ 8\ 8\ 7\ 8\ 8\ 8\ 8\ 7 \tag{4.7}$$

Consequently, consecutive samples can be accumulated into batched accordingly to Equation 4.6, as given in Equation 4.8.

$$P(j) = \sum_{n=1}^{P_L(j)} Z\left(n + \text{round}(j \cdot \overline{P})\right) \qquad (4.8)$$

$$\text{where } Z \in \{I, Q\}$$

Each partial sum $P(j)$ is multiplied with the corresponding code value j in order to form a partial correlation. Consecutive partial correlations can be combined and summed up in order to form a carrier batch $B(k)$, accordingly to Equation 4.3. However, as the ratio of partial correlations per carrier batch may result in a fractional number, it has to be rounded to the nearest integer.

The time delay between the code replicas can be achieved by shifting the sum boundaries of the partial sum in Equation 4.6 by Δn samples. This process is illustrated in Figure 4.7 with the computation of several partial sums $P(\Delta n, j)$ sequences associated to different samples offsets Δn and chips j.

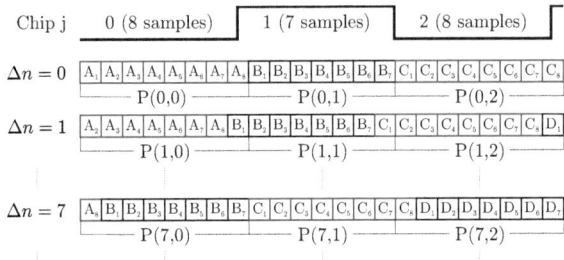

FIGURE 4.7: Example of partial sum computations

The partial sums of Figure 4.7 can be combined into a complex matrix (for both I and Q) with each row associated to a sample offset Δn and each column associated to a code chip j. The matrix U is shown in Table 4.6, where $\sum[a:b]$ denotes the partial sum from sample a to sample b.

		Chip number j			
	0 (8 samples)	1 (7 samples)	2 (8 samples)	3 (8 samples)	...
Sample offset Δn 0	$\sum[0:7]$	$\sum[8:14]$	$\sum[15:22]$	$\sum[23:30]$	...
1	$\sum[1:8]$	$\sum[9:15]$	$\sum[16:23]$	$\sum[24:31]$	...
⋮	⋮	⋮	⋮	⋮	...
7	$\sum[7:14]$	$\sum[15:21]$	$\sum[22:29]$	$\sum[30:37]$	...
⋮	⋮	⋮	⋮	⋮	...

TABLE 4.6: Example of the partial sum matrix U

Chapter 4 New architecture and algorithms for a software receiver

The exact sequence of the partial sums depends on the corresponding satellite that is processed by the channel as the PRN codes are not synchronized. This means that the matrix has to be re-calculated for each channel separately.

In order to simplify the computation process, the size of the partial sums is assumed to be constant by accumulating systematically 8 samples. This can be done by introducing a supplementary 8th sample into every sum that is originally composed of 7 samples. As this particular case intervenes every 5th or 6th chip, it corresponds to integrating the same sample twice roughly every 43rd sample.

Under this assumption, Table 4.6 can be transformed into a simplified partial sum matrix U', as shown in Table 4.7.

		Chip number j			
Sample offset Δn		0 (8 samples)	1 (7 samples)	2 (8 samples)	3 (8 samples)	
	0	$\sum[0:7]$	$\sum[7:14]$	$\sum[15:22]$	$\sum[23:30]$	...
	1	$\sum[1:8]$	$\sum[8:15]$	$\sum[16:23]$	$\sum[24:31]$	...
	⋮	⋮	⋮	⋮	⋮	...
	7	$\sum[7:14]$	$\sum[14:21]$	$\sum[22:29]$	$\sum[30:37]$	...
	⋮	⋮	⋮	⋮	⋮	...

TABLE 4.7: Example of the simplified partial sum matrix U'. The samples in red are integrated twice.

As the partial sum size is now constant, the matrix U' contains the same partial sum (e.g., $\sum[7:14]$) at two different rows. Therefore, it can be further simplified and stored as a single vector V containing all consecutive partial sums, as given in Table 4.8.

Sample offset Δn	Partial sum vector V
0	$\sum[0:7]$
1	$\sum[1:8]$
⋮	⋮
7	$\sum[7:14]$
8	$\sum[8:15]$
⋮	⋮
$N_s - 1$	$\sum[N_s : N_s + 7]$

TABLE 4.8: Example of the simplified partial sum vector V

The final sequence of the partial sums associated to the shift Δn can now be obtained accordingly to Equation 4.6 by addressing the complex partial sum vector V as given in Equation 4.9.

$$P(\Delta n, j) = V\left[\lfloor(\Delta n + j \cdot \overline{P})\rfloor\right] \tag{4.9}$$

The complex partial sum vector V can be computed iteratively with only one addition and one subtraction (except for the first sum that requires 8 additions), concurrently for all channels as given in Equation 4.10 (the same equation applies for the computation of the partial sum $P_Q(n)$). The vector must be calculated for the I and the Q branch independently.

$$\begin{aligned} P_I(0) &= \sum_{n=0}^{7}(I(n) - 4) \\ P_I(n+1) &= P_I(n) + I(n+8) - I(n) \quad &\text{for } n \in [0, N_s - 9] \\ P_I(n) &= 0 \quad &\text{for } n \geq N_s - 8 \end{aligned} \tag{4.10}$$

4.4.4 Parallel frequency search

To increase the speed of the acquisition process, the parallel frequency (post-correlation FFT) acquisition method will be implemented. This algorithm searches the correlation peak in the frequency domain by testing all the Doppler bins at once and all the code phases individually. The algorithm consists (as described in detail in Section 2.4.2 and Section 3.7.6) in forming P consecutive partial correlations with a pre-detection time T_{coh} which is P times smaller than the integration time T_{int}. The P correlation results are then re-grouped in a vector on which a N-point FFT is computed where $N \geq P$ (if $N > P$, zero padding must be applied). Assuming P large enough, all Doppler bins are searched in parallel for a given code phase. If no correlation peak is detected, the operation is repeated with the next code phase. The architecture is shown in Figure 4.8.

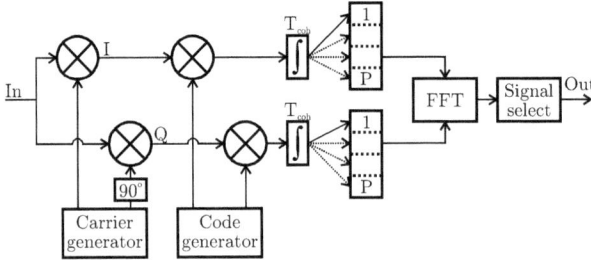

FIGURE 4.8: Parallel frequency search acquisition architecture

The architecture is characterized by the parameters given in Table 4.9.

Parameter	Description
$P \cdot T_{coh}$	Total coherent integration time T_{int}
$1 / T_{coh}$	New sampling rate and search bandwidth of the FFT
$1 / P \cdot T_{coh}$	Frequency bin resolution (bin width)

TABLE 4.9: Parameters for parallel frequency architecture

Each point of the FFT is actually a detector for the peak of the correlation for a given residual Doppler frequency. Consequently, the sensitivity of the detection is thus altered by two parameters: the integration time T_{int} leads to a global power envelope proportional to $sinc^2(\pi \cdot \Delta F_{dop} \cdot T_{coh})$, while the number of different bins leads to an envelope proportional to $sinc^2(\pi \cdot \Delta F_{dop} \cdot T_{int})$. This comportment is illustrated in Figure 4.9.

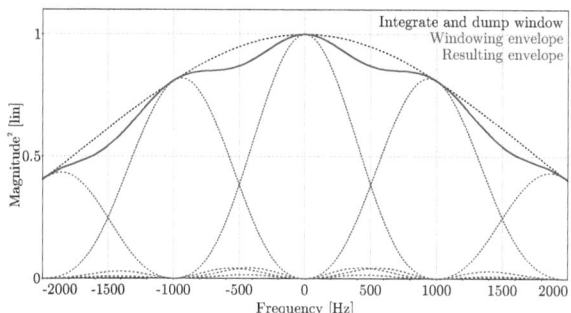

FIGURE 4.9: Frequency domain sensitivity of 4-point FFT detector

The consequences are:

- The overall sensitivity reduces when the Doppler mismatch increases. The worst loss occurs for $\Delta F_{dop} = 1/(2 \cdot T_{coh})$ and results in an amplitude of $\pi/2$.
- If the current Doppler shift falls between two Doppler bins, the sensitivity is even more reduced and two consecutive bins fire a response. However, responses of neighbor bins can be combined in order to limit this sensitivity loss.

The data of the C/A code can change during an integration period, resulting in a destructive interference. To overcome this issue, the *alternate half-bits method* is used. This consists in integrating twice on two consecutive 10 ms sequences. This way, at least one set of data is free of code change.

The structure of the partial sum vector can be fully reused for the implementation of the post-correlation FFT acquisition method. Consecutive partial sums (selected from the partial sum vector accordingly to Equation 4.9) can be regrouped and accumulated in order to recreate the P pre-detection sums on which the FFT is performed. Note that the ratio between the pre-detection size and the partial sum size may result in a non-integer value that has to be rounded to the nearest integer.

4.4.5 Parallel code search

The parallel code space acquisition tests all the code phases in parallel for a given Doppler frequency. The input baseband signal is first multiplied with the in-phase and quadrature signals and the result is transformed into the frequency domain by the mean of a FFT. The FFT of the locally generated PRN is also computed. After the multiplication of these two sets of coefficients, the inverse FFT is performed to determine the correlation peak. If no correlation peak was found, the operation is repeated with the next Doppler bin. The process is illustrated in Figure 4.10.

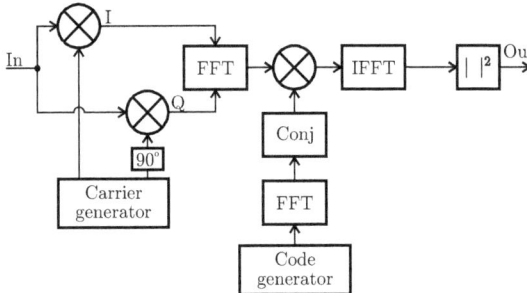

FIGURE 4.10: Parallel code search acquisition architecture

Compared to the previous method, the parallel code phase search acquisition reduces the search space to the different carrier frequencies. The Fourier transform of the locally generated PRN codes can be pre-computed and stored (assuming that the Doppler frequency on the code can be neglected) and each of the frequency bin search consists in performing one forward and one inverse Fourier transform. The number of samples in a period of the PRN code is normally not a power of 2 and the input signals must be carefully arranged with zero padding. the number of points N for the FFT computation can be expressed as given in Equation 4.11.

$$N = 2^{\lceil \log_2 (2 \cdot F_s \cdot T_{int}) \rceil} \qquad (4.11)$$

Furthermore, due to the unknown code phase of the incoming signal, one additional code period is needed to discard the end effect as illustrated in Figure 4.11.

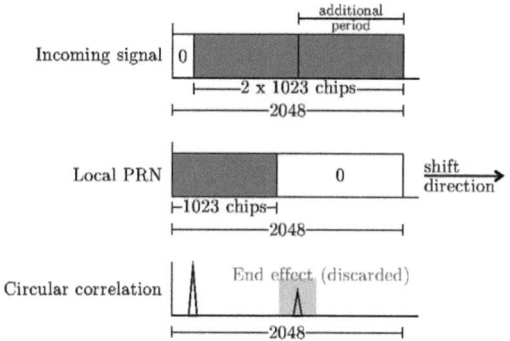

FIGURE 4.11: Circular correlation with discarded end effect

The structure of the partial sum vector (as shown in Table 4.8) can be fully reused for the implementation of the parallel code search acquisition. Consecutive partial sums (selected from the partial sum vector accordingly to Equation 4.9) are multiplied point by point with the carrier in order to remove the residual Doppler frequency. The sequence is then zero padded in order to form a vector on which the FFT is performed. The result is then multiplied point by point (in the frequency domain) with the FFT of one single PRN code period (also zero padded) in order to form a vector on which the inverse FFT is performed. The energy of each data point is computed and compared to the threshold to declare the satellite as present or not.

4.4.6 Final software receiver architecture

The use of the batch processing for both carrier and code removal leads to a redistribution of the baseband architecture, as shown in Figure 4.12.

The incoming data stream $I(n)$ and $Q(n)$ are first pre-accumulated into partial sums P_L with a length of 8 samples. This operation is performed directly on the incoming data stream and is valid for all satellites. This decreases the data rate from F_s to F_{code}. The respective partial sums sequences Early, Prompt, and Late are selected and mixed with the code at the code frequency rate before being accumulated into carrier batches B_L. This step reduces the data rate further from F_{code} to F_{dop}. Afterwards, the carrier removal can be performed at the Doppler data rate and the results are finally summed up to form the definitive accumulation results E_I, P_I, L_I and E_Q, P_Q, L_Q.

The baseband operations can be divided into two loads:

- A basic load that is common to all channels: computation of the inital partial sums;
- A load that is specific for every channel: removal of the corresponding PRN code and the residual Doppler frequency.

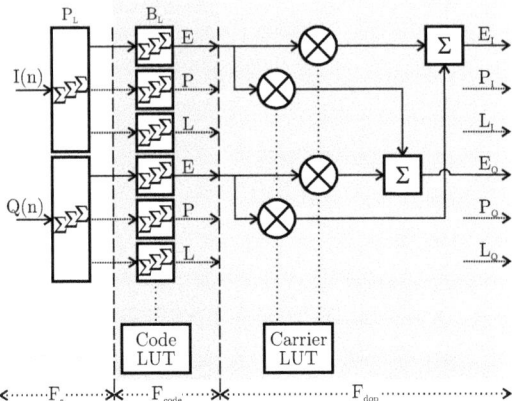

FIGURE 4.12: New software receiver baseband architecture with code and carrier batch processing

The computational load or the number of integer operations of the whole architecture is given in Table 4.10.

	Additions	Multiplications
Partial sum	$4 \cdot F_s$	0
Code generation	$N_{ch} \cdot F_{code}$	0
Code mixing	0	$6 \cdot N_{ch} \cdot F_{code}$
Carrier gen.	$N_{ch} \cdot F_{dop} \cdot 2^{N_q+1}$	0
Carrier mixing	$2 \cdot N_{cor} \cdot N_{ch} \cdot F_{dop} \cdot 2^{N_q+1}$	$4 \cdot N_{cor} \cdot N_{ch} \cdot F_{dop} \cdot 2^{N_q+1}$
Accumulation	$2 \cdot N_{cor} \cdot N_{ch} \cdot F_{dop}$	0

TABLE 4.10: Number of required operations (per second) for the new baseband architecture

The numerical values of the different operations (with the same parameters as in Section 3.7.7 and for a sampling frequency of 8 MHz) are given in Table 4.11 (for $F_s = 4$ MHz) and in Table 4.12 (for $F_s = 8$ MHz). The sampling frequency of 8 MHz is shown as the final implementation (presented in Chapter 5) will employ this sampling frequency. For the reader's convenience, the used parameters and the values from Section 3.7.7 are repeated. It is important to note that the number of needed operations for the three standard acquisitions methods does not contain the code and carrier generation!

Figure 4.13 presents a visual comparison (in a logarithmic scale!).

	Used parameters	
Sampling frequency (F_s)	4 MHz	8 MHz
Number of channels (N_{ch})	12	
Code frequency (F_{code})	1.023 MHz	
Doppler frequency range (F_{dop})	± 5 kHz	
Quantization bits (N_q)	3	
Integration time (T_{int})	10 ms	

	Additions	Multiplications
New architecture	$3.54 \cdot 10^5$	$8.52 \cdot 10^5$
without generation	$2.21 \cdot 10^5$	$8.52 \cdot 10^5$
Serial search *	$8.45 \cdot 10^{11}$	$1.69 \cdot 10^{12}$
Parallel code search *	$3.53 \cdot 10^9$	$3.64 \cdot 10^9$
Parallel frequency search *	$9.91 \cdot 10^8$	$1.97 \cdot 10^9$

* without code and carrier generation

TABLE 4.11: Comparison of numerical values of required operations for standard and new architectures (for $T_{int} = 10$ ms and $F_s = 4$ MHz)

	Additions	Multiplications
New architecture	$5.14 \cdot 10^5$	$8.52 \cdot 10^5$
without generation	$3.81 \cdot 10^5$	$8.52 \cdot 10^5$
Serial search *	$3.38 \cdot 10^{12}$	$6.76 \cdot 10^{12}$
Parallel code search *	$3.53 \cdot 10^9$	$3.64 \cdot 10^9$
Parallel frequency search *	$1.97 \cdot 10^9$	$3.93 \cdot 10^9$

* without code and carrier generation

TABLE 4.12: Comparison of numerical values of required operations for standard and new architectures (for $T_{int} = 10$ ms and $F_s = 8$ MHz)

The computation of the partial sums is performed iteratively on each incoming sample (both on I and Q), with one addition and one subtraction, except for the first partial sum where 8 additions have to be calculated.

The code generation consists in computing the number of samples in each chip (accordingly to Equation 4.9) in order to address the partial sum vector and to select the appropriate Early, Prompt, and Late sequence. This is done by the mean of a NCO updated at the code frequency rate with a phase increment equal to the mean number of samples per chip.

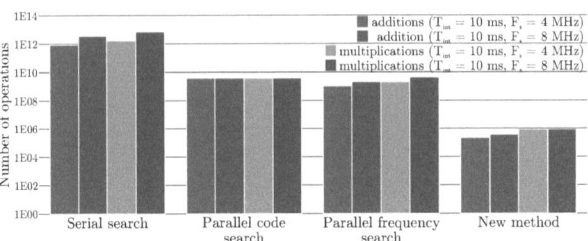

FIGURE 4.13: Visual comparison of required operations for standard and new baseband architectures (in logarithmic scale)

The code mixing is performed by multiplying each Early, Prompt, and Late partial sum sequence with the nominal chip sequence at the code frequency rate.

The carrier generation consists in computing the number of partial sums contained in each carrier batch (accordingly to Equation 4.3). This is done by the mean of a NCO updated at the Doppler frequency rate with a phase increment equal to the mean number of partial sums per batch.

The batches associated to the same phase value (i.e., every 2^{N+1th} batch) are re-grouped and accumulated to be multiplied once with the corresponding complex amplitude value.

The accumulation of the partial sums is distributed over the different carrier batches.

This new architecture offers several advantages:

- The batch processing reduces significantly the complexity (or the computational load) of the baseband operations, especially in terms of multiplications. This leads to an amount of required operations that is fully suitable for a real-time implementation even on low-cost CPUs.
- The use of a NCO for both the carrier and the code generation allows the creation of the waveforms at any desired phase and frequency, as in a standard hardware receiver. This allows a real-time compensation of the residual Doppler (on both the carrier and the code) and any quartz offset or drift. It also greatly simplifies the integration of standard tracking algorithms and loops in the receiver.
- The architecture is flexible and can easily accommodate various signal configurations (sampling frequency, signal and carrier quantization, new signals, ...) without any significant code change. Furthermore, the receiver can handle high sampling frequencies as they only affect the computation of the first partial sums in a linear way with a minimum impact on the overall complexity.
- The memory requirements are drastically reduced as no oversampled carrier or code sequence need to be stored.
- The partial sum architecture can be fully reused by the acquisition, as well as by the tracking engine.

But there are also some drawbacks:

- The Early, Prompt, and Late code replica spacing (or the correlator spacing) is restricted to integer multiples of one sample (i.e., the minimum spacing between Early-Prompt and Prompt-Late is one sample).
- The assumption of constant partial sum size (in this case: eight samples) causes a slight deformation of the correlation peak. However, this affects all satellites in the same way and does not corrupt the calculation of the code phases. This issue could also be removed by interpolating or correcting the correlation triangle.
- The carrier batch size is truncated due to the non-integer ratio between the original batch size and the partial sum size. However, the maximum error introduced is smaller than eight samples (to be compared to batch sizes in the range of [8000:100] samples).
- The solution is optimized for GNSS signals with relatively low code chip rates (like the GPS L1 with 1.023 MHz). The number of samples per partial sum decreases when the code chip rate increases, making the architecture less efficient.

This new architecture was published in a pending patent application: [62] and [63].

4.5 Summary

This chapter presented the new architecture of a real-time capable software receiver which implementation is described in Chapter 5.

The first step consisted in finding the separation between hard- and software, i.e., which part must still be present in hardware (like the antenna and the ADC) and which can be performed in software. To relax the constraints on the software side, an additional pre-processing stage was introduced (called BBPP) that converted the incoming sample stream to an appropriate configuration and format for the software receiver. The introduction of the BBPP stage allowed the development of the following architecture, as the IF (imposed by the RF front-end) was already removed and the software receiver only had to generate the Doppler frequency, normally much lower than the IF (several kHz ↔ several MHz). The incoming samples have also been arranged in an optimal way for the transmission to the software processing unit running on the host system.

Following this new configuration, the architecture for the software baseband processing units was developed and the corresponding algorithms elaborated. The main contribution was the introduction of the batch processing (for the code and carrier removal) that allowed to reduce the number of needed operations. The following sections of the chapter described the different processing blocks, like the acquisition and tracking architecture, optimized for the use with the BBPP stage.

The chapter concluded with a comparison of the new architecture with the "standard" architectures in terms of needed operations. It was shown that the new architecture reduced drastically the number of required operations, either for the baseband operations (like acquisition or tracking), as well as also for the (real-time) generation of the code and the carrier.

This chapter contains the following contributions worked out and partially published during the elaboration of this thesis:

- The proposition of using a BBPP stage to pre-process the signal allowing to lower the computational load of the software receiver;
- The main and the most important contribution elaborated during the duration of the thesis is the proposition of the completely new concept of using batch processing for the carrier and code removal. This concept also implies the inversion of the standard processing stage of carrier and code removal.

Chapter 5

Implementation of new architecture and algorithms

5.1 Introduction

This section describes the implementation of the new architecture. First, the solution on the front-end unit is presented, together with the configuration of the USB 2.0 interface. Afterwards, the implementation of the new architecture (with the use of partial sums) is detailed, together with the needed discriminator loops and a description of the realized aiding solution. The chapter closes with a schematic overview of the final software receiver prototype and an overview of the external libraries and header files.

5.2 Front-end unit

The main goal of the front-end unit is to sample, pre-process, and transmit the incoming satellites signals to the software receiver. This section covers the implementation on the front-end unit that represents the hardware part of the software receiver. The front-end unit (see Figure 5.1) consists mainly of the following components:

- RF front-end board (made with discrete components);
- FPGA (Altera Cyclone II) on FPGA-DEV board;
- USB 2.0 Controller (Cypress EZ-USB FX2LP) on FPGA-DEV board.

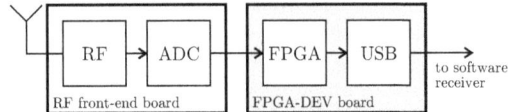

FIGURE 5.1: Front-end unit

5.2.1 RF front-end board

The RF front-end board needs a clock input for the on-board PLL (to down-convert the incoming signal) and for the ADC (to sample the down-converted signal). This clock is generated by the FPGA at a frequency of 24 MHz and the PLL is configured to generate a frequency of $F_{LO} = 1572$ MHz, resulting in an intermediate frequency of $F_{if} = 3.42$ MHz.

The summary of the specifications of the RF front-end board in given in Table 5.1.

Parameter	Value
Total power gain	102 dB
Noise figure	3 dB
N_q	5 bits
F_{LO}	1572.00 MHz
F_{if}	3.42 MHz
F_s	24 MHz

TABLE 5.1: Specification of the RF front-end board

5.2.2 FPGA

The main goal of the FPGA is programming the discrete components on the RF board (mainly the PLL and the Programmable Gain Amplifier (PGA)), pre-processing the incoming samples, and re-arranging the data for the transmission to the USB controller.

RF board programming

As written above, the FPGA is responsible for programming the PLL and PGA on the RF front-end board, with the specifications given in Table 5.1. The complete programming routine is automatically executed after the configuration of the FPGA. There is currently no automatic check implemented for testing if the programming was successful. Therefore, an external verification is necessary and recommended.

Baseband pre-processing (BBPP) unit

The BBPP unit is a top level component common to all channels. It is responsible for the following three main operations:

1. Data filtering
2. Data baseband down-conversion
3. Data bandwidth reduction

The incoming analog satellite signal is sampled at 24 MHz with a 5 bit resolution by the RF board and enters then the BBPP unit. It is first high-pass filtered (with a first order Infinite Impulse Response (IIR) filter) to remove any residual DC components and then down-converted to baseband with the help of a 3.42 MHz LO (implemented as a NCO with a resolution of 32 bits). The signal quantification is also decreased from 5 bits to 3 bits. The baseband signal is then low-pass filtered (with a Finite Impulse Response (FIR) filter of order 23 with 24 symmetrical coefficients) in order to reduce the sampling frequency from 24 MHz to 8 MHz. The signal samples are re-arranged and transmitted to the USB controller that transmits the packets to the host computer. The BBPP unit scheme is shown in Figure 5.2.

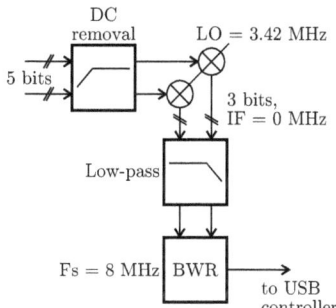

FIGURE 5.2: Baseband Pre-Processing unit

Interface to USB controller

The BBPP unit provides a data stream of $2 \cdot 3$ bits (I and Q) sampled at 8 MHz. The FPGA is in charge of re-arranging the data so that it can be transmitted as efficiently as possible to the USB controller. The interface of the Cypress USB controller is programmed to accept 16 bits in parallel. Therefore, the incoming samples (I and Q) are zero-padded to 4 bits and concatenated into an 8 bit vector, as shown in Table 5.2.

The so formed 8 bit vectors are sequentially stored in a 32 bit circular buffer at a rate of 24 MHz. When 16 bits of data are available, the content is handed to the USB controller.

D_7	D_6	D_5	D_4	D_3	D_2	D_1	D_0
0	I_2	I_1	I_0	0	Q_2	Q_1	Q_0

TABLE 5.2: Bit alignment/re-arrangement for USB interface

5.2.3 USB interface

The Cypress USB controller is the gateway between the front-end unit and the host system on which the software receiver is running, as depicted in Figure 5.3.

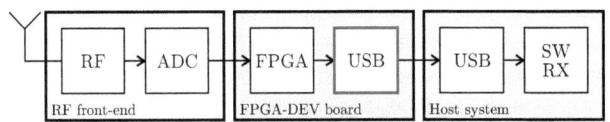

FIGURE 5.3: Position of Cypress USB controller in the complete setup

Cypress USB endpoint configuration

The USB host communicates with the connected devices through different endpoints, as described in Section 4.3.3. The following endpoints have been implemented in the Cypress USB controller for the use with the software receiver (see Figure 5.4):

o *Endpoint 1*
As the FPGA has to be programmed at every startup, a dedicated endpoint is used for this task. This is done by the use of endpoint 1 (direction *out*, mode *bulk*).

o *Endpoint 8*
As the FPGA must be able to be configured during run-time, a dedicated endpoint is used for this task. This is done by the use of endpoint 8 (direction *out*, mode *bulk*, double buffered).
The FPGA reads regularly the content of this endpoint and updates the internal registers accordingly.

o *Endpoint 2*
The incoming data has to be transmitted to the host system without losing any samples. An isochronous endpoint has a specific reserved bandwidth for data and is used for streaming the data. Endpoint 2 is defined for this task (direction *in*, mode *isochronous*, triple buffered).

The CPU clock of the Cypress USB controller is set to 48 MHz to assure a sufficient handling speed for the incoming data. The data stream from the FPGA is directly written into the corresponding endpoint FIFOs, without any interaction of the CPU. This allows a high-speed transmission of the data.

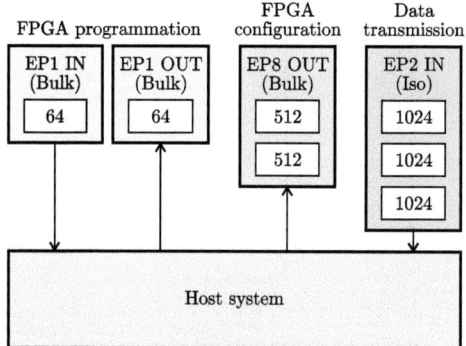

FIGURE 5.4: Configuration of USB endpoints

The Cypress USB controller is configured in a *slave FIFO* mode. That allows the FPGA to act as the master, i.e., the FPGA generates the clock for sending and receiving data to and from the Cypress USB endpoint FIFOs.

The isochronous endpoint 2 is used to get the sampled data from the front-end unit into the computer with the software receiver. To be able to transmit 8'000 samples/second, the endpoint is configured to use *triple buffering* with a packet size of 1'024 bytes. One packet is transmitted per microframe and the slave FIFO is configured with a 16 bit width to forward automatically 1'024 bytes.

With these settings, the USB controller is able to transmit data at a rate of 8 MBytes/s in isochronous mode. As the RF board outputs only 8 Msamples/s (or 8'000'000 Bytes/s), it sometimes occurs that the FIFO buffers are not completely filled with data and therefore zero-packets are transmitted. These zero-packets have to be filtered out by the software running on the computer. Details can be found in the Section 5.3.

5.3 USB real-time data handling

This section describes the USB real-time data handling on the host system.

The RF board samples the incoming GNSS signal at 24 MHz and the BBPP stage reduces the data bandwidth to 8 MHz. The samples for I and Q are each represented on 3 bits (as unsigned integers) and the FPGA aligns these values to form a 8 bit value (the remaining 2 bits are filled with 0) (see Table 5.2).

The maximum throughput from the front-end board is therefore 8 Msamples/s (or 8'000'000 bytes/s). As a continuous data stream is essential to keep the synchronization between the receiver and the respective satellite signals, great care has to be given to an uninterrupted handling of the incoming data.

The Cypress USB controller supports this throughput without any problems. USB 2.0 devices can communicate at a speed of up to 480 MBit/s (= 60 MBytes/s) in bulk mode (theoretical value), while the maximum bandwidth with isochronous transfer is 24 MBytes/s [64].

Isochronous data is time-critical and is used to *stream* data like in audio and video applications. For this reason, the isochronous transfer is chosen for the software receiver implementation. The firmware of the Cypress USB controller is described in detail in Section 5.2.3. The current section only covers the implementation on the computer side.

5.3.1 Finding device

The USB driver sets a specific GUID for the front-end unit that allows finding the device on the USB bus with the Windows function *SetupDiGetClassDivs()*. If the device and details about the device interface are found, a valid handle is returned that can be used by the software receiver to communicate with the front-end device.

This solution allows that the software receiver front-end can be connected to any available USB port and the software finds automatically the position on the USB tree.

5.3.2 Initialization

The USB interface has to be initialized correctly by the software running on the computer. Several parameters have to be set, among others:

- Number of isochronous packets per transfer;
- Length of isochronous packets;
- One or several data buffers for saving the incoming data stream;
- Special functions (or threads) to handle the data saving;
- Special events to indicate that new data is available for processing.

The number of isochronous USB packets per transfer specifies the number of packets that are requested on every transfer and has to be a multiple of 8 (as defined in the USB specifications [64]). This is due to the fact that every millisecond, eight packets are transferred what defines the length of one transfer request.

The length of the isochronous USB packets specifies how many bytes one packet contains. This value has to correspond to the value used in the USB firmware description. The default value is 1024 bytes that was also used in the following implementation.

Chapter 5 Implementation of new architecture and algorithms 99

To assure an uninterrupted transfer of the incoming data, two buffers are allocated for saving the incoming data stream (double buffering technique). Additionally, two dedicated threads are created for the USB transfer. First, a *producer thread* with high priority that handles the reading of the USB data stream from the front-end unit and the correct saving of the data into the two data buffers. Second, a *consumer thread* for getting the elapsed time information and indicating the availability of new data to the main application via events.

5.3.3 Events

When designing a real-time capable software receiver, the data processing has to be faster than the arrival of new data. If this condition is not satisfied, the receiver will not work in real-time. This implies that the receiver will have to wait for new data after having done all the calculations with the current data set. *Events* are used to indicate that new data is available. Compared to interrupts, events are normally handled synchronously, i.e., the program explicitly waits for an event to be serviced whereas an interrupt can demand service at any time.

Two events are created for the USB transfer functions:

- The *usbWriteEvent* to indicate that a new data buffer is completely filled and available for processing (watched by the consumer thread). This event is reset automatically as soon as it was detected by the consumer thread.
- The *processDataEvent* to indicate that new data is available for processing (watched by the main application). This event has to be reset manually what is done by the main application as soon as it has finished to process the data in the buffer.

This configuration allows the main application to process the data in the buffer for a longer time than the duration of the complete data buffer, e.g., during the acquisition phase where real-time data processing is not absolutely mandatory. At the same time, it assures that – as soon as the processing is finished – new data is available.

5.3.4 Threads

In software, it is not possible to start several tasks in parallel (contrarily to hardware where different tasks can run concurrently). The executed program (also called a *process*) runs sequentially and a new process can only be launched when the last one has finished. Nevertheless, modern operating systems offer the possibility to run several tasks in parallel. On a single processor, multithreading generally occurs by time-division multiplexing and the processor switches between the different threads. This context switching generally happens frequently enough that the user perceives the threads or tasks as running at the same time (the Windows operating system switches between the different tasks every

10 – 15 ms [65]). On a multiprocessor or multi-core system, the threads or tasks will generally run at the same time, with each processor or core running a particular thread or task.

In a software receiver, the digitized samples are streamed from the ADC (and in this case, pre-processed by the FPGA) to the host system. An interruption of the incoming data stream would lead to a sudden loss of tracking and the satellites would have to be re-acquired again. Therefore, special care has to be taken that no data samples are lost during transmission and that the software is not occupied with another task and can therefore not read out new data. To assure an uninterrupted transfer, two additional threads (beside the main application) are created:

o The *producer thread* for reading and saving the incoming data stream;
o The *consumer thread* for incrementing the millisecond and 20 milliseconds counters and indicating the availability of new data to the main application.

Producer thread

The producer thread initializes several (typically 8 to 16) concurrent USB transfer requests to assure an uninterrupted transfer. As only one transfer request can be active at the same time, the remaining transfer requests are queued automatically by the operating system. As soon as the first transfer finishes, the second is executed and the data of the first transfer is read. The number of received data is compared to the defined USB transfer length to check if a block with empty data was transmitted. The empty data block is removed by slicing the complete received data block to the smallest element (one frame = 1'024 bytes) and comparing them with a zero block. The data containing information is handed over to the function that saves the data in the correct data buffer. The transfer that just finished is reset and re-launched again (i.e. it is queued by the operating system as another transfer is currently active).

There are two data buffers implemented, each containing 160'000 samples (or bytes) that corresponds to 20 ms of data. The function responsible for saving the data into the correct data buffer, obtains the new data from the producer thread and checks if there is enough space in the currently active data buffer to save the complete packet. If this is possible, the new data is entirely placed in the data buffer and the writing pointer is incremented by the length of the packet. If there is not enough space for the whole packet, the current active data buffer is filled up and the remaining bytes are written to the second data buffer that becomes the active data buffer for the subsequent transfers. To signalize a complete buffer, an event is generated (named *usbWriteEvent*) that is permanently watched by the consumer thread.

Consumer thread

The consumer thread waits for the *usbWriteEvent* and as soon as this event is set, the thread copies the new data buffer into a temporary buffer that will be processed by the main application afterwards, increments the 1ms and the 20ms counter and creates an event for the main application (*processDataEvent*) to indicate that the data is ready to be processed. The *usbWriteEvent* is reset automatically.

The consumer thread increments the two counters every time a new data buffer is available (i.e., every 20 milliseconds). The 1ms counter is incremented by 20 while the 20ms counter is incremented by one. This gives in fact two parallel counters with a scaling factor of 20. While the 20ms counter indicates the number of processed data buffers, the 1ms counter is used as the internal time base and assures that the software receiver gets accurate information of the elapsed time, even if the main application did not yet finished to process the data. It is more accurate to count the number of processed samples than to rely on the internal clock of the operating system: As Windows is a non-real-time operating system, the accuracy of the internal PC clock is limited to the base of the Windows task scheduler, which is 10 ms.

The main application resets the *processDataEvent* manually as soon as it has finished. This operations assures that only the latest available data is processed. The relation between the threads is illustrated in Figure 5.5.

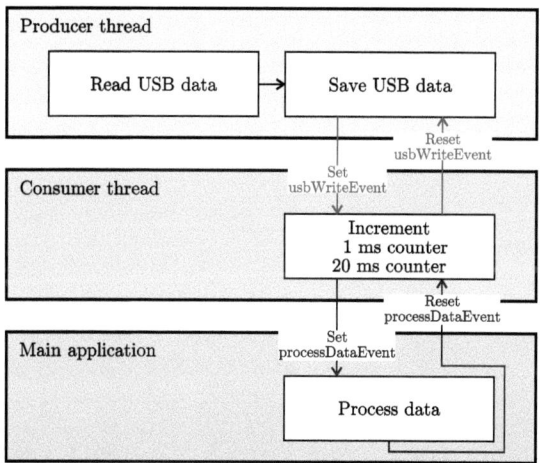

FIGURE 5.5: Threads relations and events

5.3.5 Notes

In terms of programming complexity, the isochronous transfer method is definitively not the easiest transfer mode. Its setup and the correct handling of the returns are more complicated than with the other transfer methods. When working with isochronous transfers, attention has to be paid to the expected bandwidth. It is not recommended to use the highest transfer rate because this would cause a lot of empty frames (only zero values are transmitted) that have to be filtered out after reception. Therefore, it would be perfect to use exactly the same speed as the incoming data stream. Unfortunately this is often not possible as the USB specification defines only packet sizes of 512 or 1024 bytes.

5.4 Baseband processing

This section describes the implementation of the baseband processing algorithms, as presented in Section 4.4.1.

The baseband processing can be divided in three stages: acquisition, re-acquisition, and tracking. All channels are processed together within the same stage and the receiver waits for all channels to be ready before switching to the next stage. This means that all channels are in the same state. Once in tracking stage, there is currently no way for a channel to go back into a previous stage; this makes the re-acquisition of lost satellites and the acquisition of new satellites impossible.

This section first explains the implementation of the partial sum computation. This component is the main core of the whole architecture and it is used by all other baseband processes. Then, the implementation of the code and carrier generation are presented needed to address the correct partial sums used afterwards for the code and carrier removal. Finally, the implementation of the three processing stages is presented, as well as the code phase measurements and the simple tracking loops.

5.4.1 Computation of the partial sums

The incoming analog data is sampled by the ADC on the front-end unit at a rate of 8 MHz (for I and Q). The two corresponding samples (each a 3 bit value) are aligned in the FPGA to one byte D allowing to reduce the throughput on the USB interface. The bit alignment is show in Table 5.3.

D_7	D_6	D_5	D_4	D_3	D_2	D_1	D_0
0	I_2	I_1	I_0	0	Q_2	Q_1	Q_0

TABLE 5.3: Bit alignment of transmitted data byte D

Both I and Q samples are given as unsigned integers and can be easily extracted as shown in Equation 5.1.

$$I_n = D >> 4$$
$$Q_n = D \mathbin{\&} 7$$
(5.1)

The partial sums $P_I(n)$ and $P_Q(n)$ are computed iteratively with each $I(n)$ and $Q(n)$ sample, using one addition and one subtraction, as given in Equation 4.10. Both $I(n)$ and $Q(n)$ samples are originally signed values (output from the ADCs on the RF front-end board) in the range of $[-4;3]$ and are transformed in the FPGA into unsigned integers in order to simplify their decoding in Equation 5.1. This offset is compensated in the first partial sums $P_I(0)$ and $P_Q(0)$ by subtracting the decimal value of 4 from each sample. In the subsequent calculations ($P_I(n+1)$ and $P_Q(n+1)$), this operation is no longer necessary as the offset is eliminated by the sum and the subtraction.

5.4.2 Code generation

The code generator is implemented as a standard NCO updated at the code frequency rate (i.e., no oversampled version of the code is generated). The code NCO increment is the average of the number of samples per chip ($\overline{P} = F_s/F_{code}$). The accumulated value is rounded to the next lower integer and used to address the partial sum vector in order to select the correct partial sums. The different code replicas (Early, Prompt, and Late) are obtained by shifting the original pointer by a constant offset (3 samples to obtain nearly a half chip correlator spacing). The idea is illustrated in Figure 5.6.

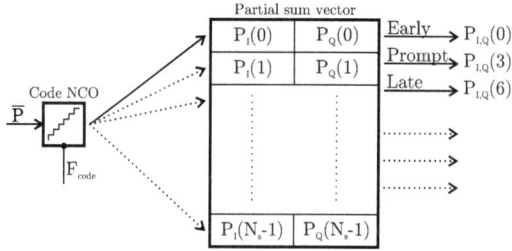

FIGURE 5.6: Code NCO used for addressing the partial sum vector V

Table 5.4 shows an example of the addressing obtained from a NCO with an average number of samples per chip of 7.82 (= increment value) and a correlator spacing of 3 samples.

		Chip					
	0	1	2	3	4	5	...
NCO accumulator	0	7.82	15.64	23.46	31.28	39.1	...
Address (Early)	0	7	15	23	31	39	...
Address (Prompt)	3	10	18	26	34	42	...
Address (Late)	6	13	21	29	37	45	...

TABLE 5.4: Addressing obtained from a NCO with a phase increment of 7.82 and a correlator spacing of 3

The encoding of the number of samples per chip (either 7 or 8) requires 4 bits ($8_{10} = 1000_2$). The increment value is multiplied by 2^{28} and stored as a 32 bit unsigned integer to get a sufficient accuracy of the addressing and to avoid floating point calculations. The 4 Most Significant Bits (MSBs) represent the integer part and the 28 Least Significant Bits (LSBs) the floating part. The same technique is also applied for the accumulator.

In the final implementation, the 4 MSBs are transferred and integrated separately into the addressing pointer of the partial sum vector and reset after each iteration to avoid an overflow of the accumulator. This approach is shown in Table 5.5. Please note the difference to Table 5.4 in the row '*NCO accumulator*'.

		Chip					
	0	1	2	3	4	5	...
NCO accumulator	0	7.82	8.64	8.46	8.28	8.1	...
Address (Early)	0	7	15	23	31	39	...
Address (Prompt)	3	10	18	26	34	42	...
Address (Late)	6	13	21	29	37	45	...

TABLE 5.5: Modified addressing obtained from a code NCO with a phase increment of 7.82 and a correlator spacing of 3

The implementation is described in Equation 5.2.

$$\begin{aligned} \text{NCO}_\text{P} &= (\text{NCO}_\text{P} \ \& \ 2^{28}) + \overline{P} \\ \text{Pointer}_\text{P} &= \text{Pointer}_\text{P} + (\text{NCO}_\text{P} >> 28) \end{aligned} \quad (5.2)$$

This modifications brings the advantage that all the operations are executed with 32 bit unsigned integers and the increment and the accumulator has a precision of 28 bits.

5.4.3 Carrier generation

The carrier generation is implemented as a standard NCO updated at the Doppler frequency rate. The increment corresponds to the average number of samples per carrier batch (i.e., $\overline{B} = F_s / \left(2^{N_q+1} \cdot F_{car}\right)$).

This means that the code and the carrier NCO are updated at different rates (code and carrier rate, respectively). In order to simplify the implementation, the code and the carrier generators are synchronized and updated at the code frequency rate (which is higher than the Doppler frequency rate). This allows integrating both operations into one single loop over the chips and avoids the overlapping of two calculations at different frequencies. Therefore, the increment of the carrier NCO changes corresponding to the number of carrier phases per chip (i.e., $\overline{B} = \left(2^{N_q+1} \cdot F_{car}\right) / F_{code}$).

The accumulator value is rounded to the next lower integer value and used to determine the carrier phase value corresponding to the current chip, as illustrated in Figure 5.7.

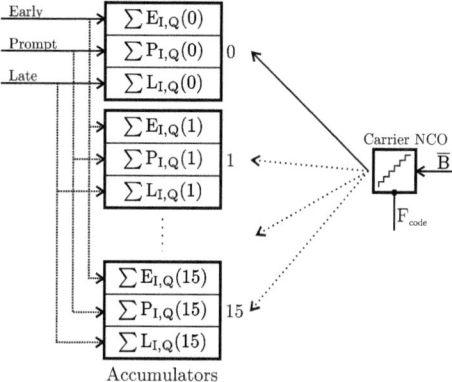

FIGURE 5.7: Carrier NCO addressing the 16 carrier batch accumulators associated to the different carrier phases

The process should be illustrated by the example of a NCO generating a carrier frequency of 5 kHz. The accumulator is incremented at the nominal code frequency rate (1.023 MHz) with the average number of carrier batches per chip equal to 0.039 (for $N_q = 2$). Each accumulator overflow represents the beginning of a new carrier cycle. The accumulator increases as shown in Table 5.6.

	Chip					
	0	1	...	25	26	...
NCO accumulator	0	0.039	...	0.977	1.017	...
Carrier batch	0	0	...	0	1	...

TABLE 5.6: Carrier NCO updated at the code frequency rate with a phase increment of 0.039 carrier batches per chip

In the final implementation, the 16 different carrier phases (for $N_q = 3$) are encoded with 4 bits. Consequently, the increment value is multiplied by 2^{28} and stored as a 32 bit unsigned integer, with the 4 MSBs representing the integer part and the 28 LSBs the floating part of the carrier phase. The same principle applies for the accumulator.

The implementation is described in Equation 5.3.

$$\text{NCO}_B = \text{NCO}_B + \overline{B}$$
$$\text{Pointer}_B = \text{NCO} >> 28$$

(5.3)

This modification brings the same advantage as for the code NCO implementation: all the operations are executed with 32 bit unsigned integers and the increment and the accumulator have a precision of 28 bits.

5.4.4 Code removal

The code removal process is performed at the code frequency rate by multiplying each partial sum with the corresponding chip value. The result is then integrated in an accumulator correspondingly to the current value of the carrier NCO.

In order to reduce the amount of operations involved in the code removal process, the partial sums are stored into two different accumulators accordingly to the current chip value (0 or 1), as illustrated in Figure 5.8.

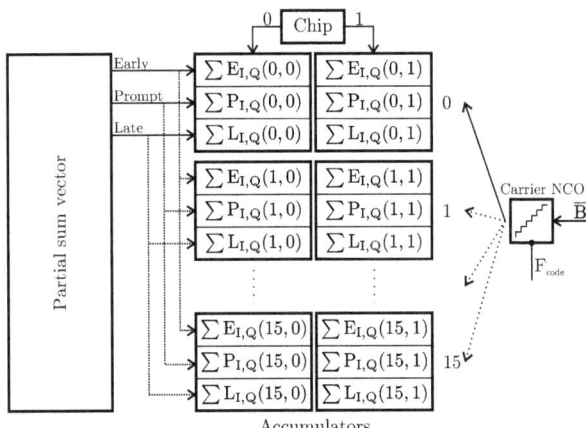

FIGURE 5.8: The 2x16 complex accumulators are addressed accordingly to the chip value and the carrier NCO

At the end of each integration period, the second accumulator result is subtracted from the first one in order to recreate the original correlation result, as given in Equation 5.4. This operation is done for every carrier phase, giving a total number of 16.

$$\sum E_{I,Q}(i) = \sum E_{I,Q}(i,1) - \sum E_{I,Q}(i,0)$$
$$\sum P_{I,Q}(i) = \sum P_{I,Q}(i,1) - \sum P_{I,Q}(i,0) \quad (5.4)$$
$$\sum L_{I,Q}(i) = \sum L_{I,Q}(i,1) - \sum L_{I,Q}(i,0) \quad \text{for } i \in [0;15]$$

As this operation runs at the frequency of the code, 16 complex accumulators are dumped and stored separately for every millisecond.

5.4.5 Carrier removal

The carrier removal consists in accumulating and multiplying the 16 complex accumulator values obtained from Section 5.4.4 with their corresponding carrier amplitude levels ($\pm 1, \pm 2, \pm 3, \pm 4, \pm 5$) in order to obtain the final in-phase and quadrature correlation values $\sum E_I, \sum P_I, \sum L_I$ and $\sum E_Q, \sum P_Q, \sum L_Q$. The carrier removal is executed every millisecond following Equation 5.5 (only the equations for the Early components are given, but the same principle applies for the Late and Prompt) with the components given in Equation 5.6 and Equation 5.7.

$$\sum E_I = 1 \cdot \left[\sum E_{II}(0) + \sum E_{QQ}(0) \right] + 2 \cdot \left[\sum E_{II}(1) + \sum E_{QQ}(1) \right] \\ + 4 \cdot \left[\sum E_{II}(2) + \sum E_{QQ}(2) \right] + 5 \cdot \left[\sum E_{II}(3) + \sum E_{QQ}(3) \right]$$

$$\sum E_Q = 1 \cdot \left[\sum E_{QI}(0) - \sum E_{IQ}(0) \right] + 2 \cdot \left[\sum E_{QI}(1) - \sum E_{IQ}(1) \right] \\ + 4 \cdot \left[\sum E_{QI}(2) - \sum E_{IQ}(2) \right] + 5 \cdot \left[\sum E_{QI}(3) - \sum E_{IQ}(3) \right]$$

(5.5)

$$\sum E_{II}(0) = \sum E_I(4) - \sum E_I(12) - \sum E_I(3) + \sum E_I(11)$$
$$\sum E_{IQ}(0) = \sum E_I(7) - \sum E_I(15) + \sum E_I(0) - \sum E_I(8)$$
$$\sum E_{II}(1) = \sum E_I(5) - \sum E_I(13) - \sum E_I(2) + \sum E_I(10)$$
$$\sum E_{IQ}(1) = \sum E_I(6) - \sum E_I(14) + \sum E_I(1) - \sum E_I(9)$$
$$\sum E_{II}(2) = \sum E_I(6) - \sum E_I(14) - \sum E_I(1) + \sum E_I(9)$$
$$\sum E_{IQ}(2) = \sum E_I(5) - \sum E_I(13) + \sum E_I(2) - \sum E_I(10)$$
$$\sum E_{II}(3) = \sum E_I(7) - \sum E_I(15) - \sum E_I(0) + \sum E_I(8)$$
$$\sum E_{IQ}(3) = \sum E_I(4) - \sum E_I(12) + \sum E_I(3) - \sum E_I(11)$$

(5.6)

$$\sum E_{QQ}(0) = \sum E_Q(7) - \sum E_Q(15) + \sum E_Q(0) - \sum E_Q(8)$$
$$\sum E_{QI}(0) = \sum E_Q(4) - \sum E_Q(12) - \sum E_Q(3) + \sum E_Q(11)$$
$$\sum E_{QQ}(1) = \sum E_Q(6) - \sum E_Q(14) + \sum E_Q(1) - \sum E_Q(9)$$
$$\sum E_{QI}(1) = \sum E_Q(5) - \sum E_Q(13) - \sum E_Q(2) + \sum E_Q(10)$$
$$\sum E_{QQ}(2) = \sum E_Q(5) - \sum E_Q(13) + \sum E_Q(2) - \sum E_Q(10)$$
$$\sum E_{QI}(2) = \sum E_Q(6) - \sum E_Q(14) - \sum E_Q(1) + \sum E_Q(9)$$
$$\sum E_{QQ}(3) = \sum E_Q(4) - \sum E_Q(12) + \sum E_Q(3) - \sum E_Q(11)$$
$$\sum E_{QI}(3) = \sum E_Q(7) - \sum E_Q(15) - \sum E_Q(0) + \sum E_Q(8)$$

(5.7)

The carrier removal results in 20 complex correlation values coherently integrated over 1 ms. Note that each code period is integrated separately in order to provide a maximum of flexibility to the further tracking loops.

5.4.6 Acquisition

The receiver has to deal with a large Doppler offset ($\pm$ 50 kHz), mainly due to the quartz offset that can take a value up to $\pm$ 45 kHz. The acquisition stage use the parallel frequency search (post-FFT) algorithm as described in Section 4.4.4 and is characterized by the numeric parameters given in Table 5.7.

Parameter	Value
Pre-detection time	$T_{coh} = 9.8\ \mu s$
Consecutive partial correlations	$P = 1024$
Total coherent integration time	$T_{int} = P \cdot T_{coh} = 10$ ms
Frequency search bandwidth	$1/T_{coh} = 102'400$ Hz
Frequency bin width	$1/T_{int} = 100$ Hz
FFT size	$N = 1024$
Correlator spacing	$1/8^{th}$ of chip

TABLE 5.7: Parameters for acquisition stage

The sensitivity of the acquisition is influenced by the following two processes: the pre-detection time leads to a global power envelope (dashed blue line in Figure 5.9) proportional to $sinc^2(\pi \cdot \Delta F \cdot T_{coh})$, while the number of bins and thus their frequency width leads to the frequency bin envelope (red line in Figure 5.9, centered at 0 Hz) proportional to $sinc^2(\pi \cdot \Delta F \cdot T_{int})$.

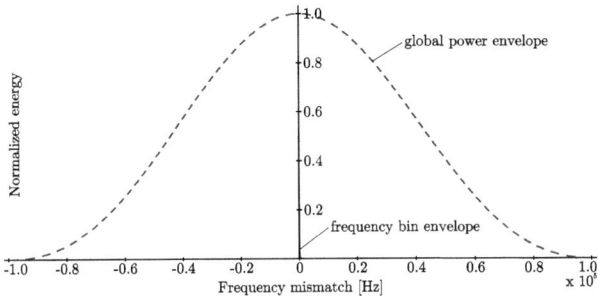

FIGURE 5.9: Acquisition sensitivity

The worst case occurs when the Doppler shift falls exactly between two neighboring bins. In this case, both bins fire a response and the sensitivity drops of up to -3.92 dB (blue line in Figure 5.10). This loss can be partially recovered by combining two adjacent frequency bins (e.g., by computing the difference between two adjacent bin responses). It is theoretically possible to recover about 3 dB and limit the degradation to approximately -1 dB (dashed red line in Figure 5.10). For more details, refer to [11].

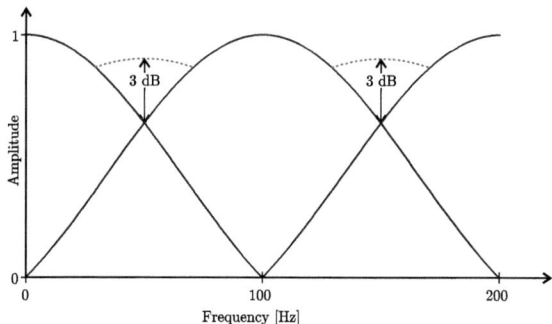

FIGURE 5.10: Recovery for post-FFT sensitivity loss

The following steps are performed in the acquisition stage:

1. Computing the partial sums of the incoming data stream by accumulating 8 samples (see Section 5.4.1);
2. Selecting a sequence of 10'240 partial sums (= 81'920 samples or 10 milliseconds) by addressing the partial sum vector with the help of the code NCO (see Section 5.4.2);
3. Multiplying the obtained sequence point by point with a sequence of 10'240 chips ($\cong$ 10 ms of code) in order to remove the PRN code;
4. Regrouping and accumulating the results into batches of approximately 100 chips in order to form the 1024 pre-detection sums;
5. Performing the FFT on the 1024 pre-detection sums;
6. Computing the energy of each data bin as well as the energy resulting from the difference of two consecutive data bins;
7. Comparing the obtained value the threshold to declare the satellite present or not.

In order to obtain a 1/8 of chip search resolution, the whole operation is repeated up to 8 times, each time selecting a new partial sum sequence shifted by 1 sample with the respect to the previous one. The implementation of the parallel frequency search exploits therefore completely the partial sum structure.

When the satellite is found in the incoming signal, the corresponding information is saved and the channel is marked to be processed afterwards by the re-aquisition stage.

5.4.7 Re-Acquisition

The re-acquisition process is implemented as a parallel code search on the frequency bins provided by the first acquisition stage (Section 5.4.6). This guarantees a short processing time as the search space is reduced from ± 45 kHz to a few frequency bins only. Additionally, possible code drifts can be detected more easily.

The term *re-acquisition* does not mean that a lost satellite can be found by this method but this stage is used to confirm the presence of the satellite in the incoming signal and to refine the Doppler frequency and the code phase. This stage is necessary because all channels are in the same state (acquisition or tracking) and a significant amount of time between the first and the last successful acquisition can elapse.

The re-acquisition stage is characterized by the numeric parameters given in Table 5.8.

Parameter	Value
Coherent integration time	$T_{int} = 7$ ms
FFT size	$N = 8192$
Correlator spacing	$1/8^{th}$ of chip

TABLE 5.8: Parameters for re-acquisition stage

The length of the input signals for the FFT calculation must be carefully selected and arranged with zero padding as the number of chips in a PRN code period is not a power of 2 ($1023 \neq 2^n$). The possible sizes for the FFT are either 8'184 or 16'368 chips (that will be zero-padded to 8'192 or 16'368 values, respectively) to provide approximately 10 ms coherent integration time. The 8'184 chips correspond to 8 ms, while the 16'368 chips correspond to 16 ms. Due to the requirements of the alternate half-bits method (consisting in integrating twice on a maximum of 10 ms to overcome a possible data bit transition), only the 8 ms chip sequence is considered. As one additional code period is needed to discard the end effect (see Section 4.4.5) caused by the unknown code phase of the incoming signal, the equivalent coherent integration time is reduced from 8 to 7 ms.

The following steps are performed in the re-acquisition stage:

1. Computing the partial sums of the incoming data stream by accumulating 8 samples (see Section 5.4.1);
2. Selecting a sequence of 8'184 partial sums by addressing the partial sum vector with the help of the code NCO (see Section 5.4.2);

3. Multiplying the obtained sequence point by point with the carrier (generated in real-time by a NCO) to remove the residual Doppler frequency (see Section 5.4.5);
4. Left zero padding of the sequence to form a vector of 8'192 points on which a FFT is performed;
5. Multiplying the obtained result point by point with the FFT of one single PRN code period (right zero padded to 8'192 points);
6. Performing the inverse FFT on the obtained sequence;
7. Computing the energy of each data point;
8. Comparing the result to the threshold to declare the satellite present or not.

As in the acquisition stage, this operation is repeated 8 times with a shifted partial sum to obtain a 1/8 of chip search resolution. The implementation of the parallel frequency search completely exploits the partial sum structure.

The codes of the acquired satellites may have changed (typically a few tens of chips) during the time spent in the re-acquisition stage due to the quartz short-term instability. Consequently, the measured code phases are extrapolated with respect to the code frequency and the re-acquisition time before entering in the tracking mode. The parallel code search function then transmits the respective satellite Doppler frequency and code offsets to the tracking stage.

The two acquisition stages use datasets of 40 ms length, processed in four packages of 10 ms. The search is initially performed on the first 10 ms sequence and – if a signal is detected (i.e., the correlation peak is higher than a pre-defined threshold) – the third 10 ms sequence is processed to confirm the signal presence. If no signal is detected, the search is done on the second 10 ms data sequence (and respectively fourth sequence for the confirmation) in order to get rid of any possible data bit transition.

5.4.8 Tracking

The tracking unit processes data sequences of a length of 20 ms, composed of $2 \cdot 160'000$ samples each. This value is chosen as a trade-off between speed and memory requirements. The first operation consists in computing the 160'000 complex partial sums for the whole data sequence. As the same vector is shared by all the tracking channels this operation is performed only once every 20 ms. All further operations are specific to every satellites and therefore need to be repeated for each channel.

The operations of the tracking stage are shown in Figure 5.11. The flow-chart is run through for every channel and every 20 ms (except the calculation of the partial sums which are only calculated once every 20 ms).

The NCO code and carrier increments are computed first as they have to be updated in order to compensate the residual Doppler frequency and the quartz offset. The next step is the correlation operation which is implemented as a loop iterated at the code frequency rate. For each chip, the following operations are performed sequentially:

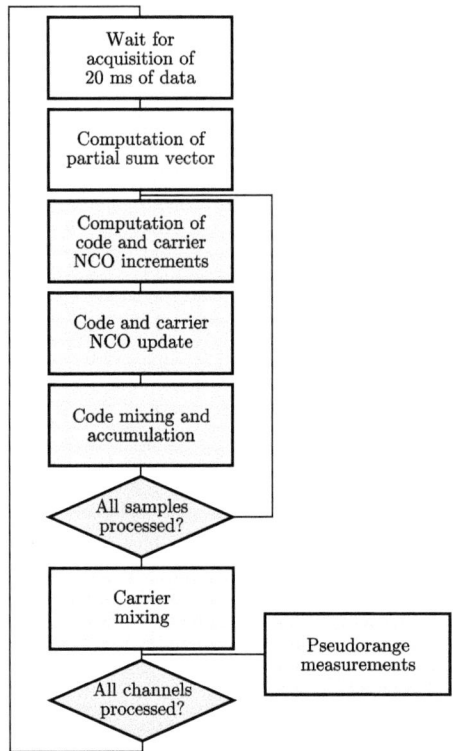

FIGURE 5.11: Flowchart of tracking unit (only one channel illustrated). The partial sum vector is computed once for all tracking channels

1. The code NCO is first updated in order to determine the number of samples contained in the current chip (7 or 8) and addresses the partial sum vector accordingly (see Section 5.4.2 and Section 5.4.4);
2. The carrier NCO is updated in order to determine the corresponding carrier phase (value between 0 and 15) associated to the current chip and addresses the appropriate carrier batch accumulator;
3. The code mixing is performed by multiplying the selected partial sums with the current chip value and by integrating the result into the accumulator selected in step 2.

As the exact number of chips in the 20 ms data sequence is generally not an integer value, it is rounded to the next higher integer (and not floored in order to guarantee that all the 160'000 samples are processed). If needed, dummy samples are used for completing the last partial sum. When the loop over the chips has finished, the carrier removal is performed in order to provide 20 complex correlation results coherently integrated over 1 ms. The code phase measurement is done and the code and carrier tracking loops are updated to determine the parameters of the next tracking iteration.

5.4.9 Code phase measurements

The code phase measurements are performed every 20 milliseconds, i.e., the measurements are always synchronized with the $160'000^{th}$ sample for each channel. As the number of samples processed $N_{samples}$ is generally slightly higher than 160'000, interpolation is needed to obtain the exact code phase measurement. A typical example of the code phase measurement is illustrated in Figure 5.12.

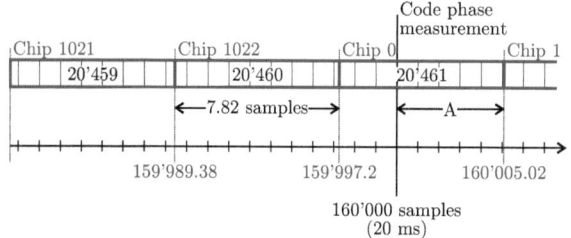

FIGURE 5.12: Code phase measurement

In the above example, the main tracking loop is iterated 20'461 times (on a chip basis) and the exact number of processed samples $N_{samples}$ is 160'005.02. This information is stored in the code NCO and can be obtained by combining the pointer of the chip sum (= 160'005) with the floating part of the code NCO accumulator (scaled by 2^{28}) (= 0.02), as given in Equation 5.8.

$$N_{samples} = \text{Pointer}_P + \left[(\text{NCO}_B \ \& \ 2^{28}) >> 28\right]$$
$$= 160'005.02 \text{ samples} \tag{5.8}$$

As the code phase measurement has to be calculated after 160'000 samples, the unknown A in Figure 5.12 can be found by Equation 5.9.

$$A = N_{samples} - 160'000$$
$$= 5.02 \text{ samples} \tag{5.9}$$

The exact state of the chip counter after 160'000 samples can be found with Equation 5.10.

$$\text{chip counter} = 1 - \frac{A}{P}$$
$$= 1 - \frac{5.02}{7.82} = 0.3581 \text{ chips} \tag{5.10}$$

with $\overline{P}$ being the code NCO increment corresponding to the average number of samples per chip.

The measured code phases are afterwards transmitted to the navigation solution to obtain a position fix.

5.4.10 Tracking loops

The tracking loops of the receiver are kept as simple as possible to limit the computational load of the software receiver and because the main goal of the implementation is to show the performance of the new architecture.

Code loop

The code loop is implemented as a standard second order DLL, as shown in Figure 5.13 [3]. The code discriminator is a non coherent Early-minus-Late with a correlator spacing of 3 samples, computed as in Equation 5.11.

$$E = \sqrt{\sum E_I^2 + \sum E_Q^2} \qquad L = \sqrt{\sum L_I^2 + \sum L_Q^2}$$

$$DLL = \frac{E + L}{2 \cdot (E - L)}$$

(5.11)

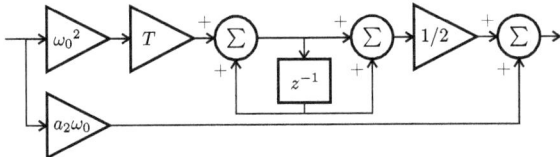

FIGURE 5.13: Second order DLL

The values $\sum E_I, \sum E_Q$ and $\sum L_I, \sum L_Q$ are obtained by non-coherently integrating all the 20 correlation results (that are coherently integrated over 1 ms) provided by the carrier removal stage. This guarantees the integrity of the DLL measurements as they are not affected by a possible data bit transition. The code loop is updated every 20 ms.

Carrier loop

The carrier loop is implemented as a second order PLL assisted by a first order FLL, as shown in Figure 5.14 [3]. If the FLL error is set to zero, the filter becomes a pure second order PLL and vice versa.

Chapter 5 Implementation of new architecture and algorithms

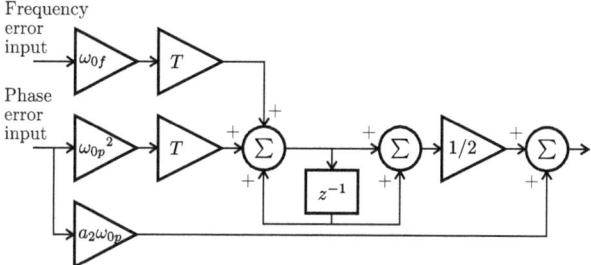

FIGURE 5.14: Second order PLL assisted by a first order FLL

The phase and frequency discriminators are implemented as described in Equation 5.12.

$$Dot = \sum P_I(t_2) \cdot \sum P_I(t_1) + \sum P_Q(t_2) \cdot \sum P_Q(t_1)$$
$$Cross = \sum P_I(t_1) \cdot \sum P_Q(t_2) - \sum P_I(t_2) \cdot \sum P_Q(t_1)$$
(5.12)

$$PLL = \frac{\sum P_Q}{\sum P_I} \qquad FLL = \frac{\text{atan}(Dot, Cross)}{t_2 - t_1}$$

In order to be able to deal with high dynamic environments (such as the quartz short term variation), a short pre-detection time is required. Therefore $\sum P_I$ and $\sum P_Q$ are coherently integrated over 2 ms, by coherently adding 2 consecutive correlation results (that are each coherently integrated over 1 ms). Several phase and frequency measurements are performed and averaged over 20 ms, minimizing the influence of a possible data bit transition. The carrier loop is also updated every 20 ms.

5.5 Aiding

The section describes the implementation of the aiding solution for the software receiver. Aiding consists in a list of visible satellites for the acquisition (to improve the acquisition time) and valid ephemeris for the navigation solution (as the navigation data is not decoded).

The navigation solution needs some basic aiding information to be able to calculate the Position, Velocity, and Time (PVT) solution. This includes the ephemeris of the tracked satellites (to be able to position the satellites in the sky) and an approximate position and time (to be able to position the receiver on the ground and in time).

The aiding can be performed by two methods:

1. Using an external aiding receiver that is connected to the same input signal (that delivers the ephemeris, position, and time);
2. Using saved aiding information (ephemeris, last known position, current time).

An external aiding receiver (in this case a u-blox 5 evaluation kit (EVK-$5H$)) is connected to the same input signal source (simulator or real signal) as the software receiver. This receiver acquires the satellites and decodes the ephemeris data of the visible satellites and performs a time synchronization. The information and the current position is polled by the software receiver and forwarded to the navigation solution that determines the list of visible satellites. This list is returned to the software receiver that begins to acquire the visible satellites.

The same information can also be provided by external files and the list of satellites to search can be specified in an external configuration file. This file determines if the software receiver uses an external receiver of saved information for the aiding. If the option for the aiding receiver is activated, the software receiver establishes a serial connection to the external receiver and polls and saves the required information. As the accuracy of the received information may be too good, the software receiver falsifies the position and decreases the accuracy of the position and of the time. If this step is not done, the navigation solution can take the (possibly more accurate) aiding information as the reference. In this case, the PVT solution is not calculated with the data from the software receiver. The modified messages are finally handed to the navigation solution.

If the option for using external files is activated (i.e., the external aiding receiver is not used), the software receivers reads the files containing the time and ephemeris information and also falsifies the values (to be sure that the navigation solution calculates a new PVT solution). It then sends the modified messages to the navigation solution which returns a list of visible satellites that have to be searched, as it is done for the configuration with an external receiver.

It is also possible to specify a default list of satellites in the external configuration file. The aiding information still have to be sent to the navigation solution, but the returned list of visible satellites is ignored and the software receiver starts acquiring the satellites given in the external configuration file.

5.6 Final software receiver prototype

Figure 5.15 shows the schematic view of the final software receiver prototype, including all the different elements needed for operation:

- The front-end unit (with the RF front-end, the FPGA, and the USB controller) for digitizing and transmitting the data;

- The host system running the software receiver and the navigation solution, together with a graphical user interface for showing the results;
- The optional external aiding receiver.

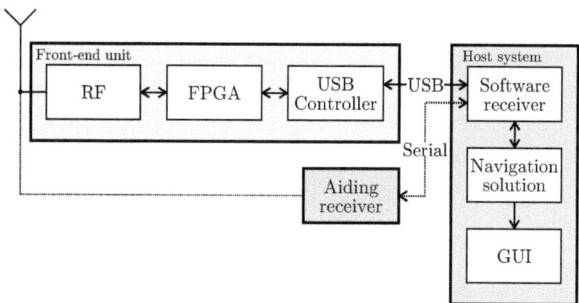

FIGURE 5.15: Final software receiver prototype

5.7 External libraries and header files

The main goal of the development is to create a software receiver that is independent of a specific CPU architecture. This implies that CPU specific commands (as SIMD instructions) have to be avoided. Nevertheless, some functions depend directly on the libraries of the operating system (as creating additional threads and events) while others make use of external libraries (as the calculation of the FFT) as they allow the fastest implementation and execution time.

This section describes the external libraries and header files that have been used for the implementation of the software receiver. It gives an overview of the changes that to be performed when the software is compiled for another platform.

5.7.1 FFTW

FFTW is a free C subroutine library for computing the discrete Fourier transform (DFT) in one or more dimensions, of arbitrary input size, and of both real and complex data. The library uses specific CPU routines as SSE/SSE2/SE3/3dNow! and supports different platforms and microprocessors. The availability of the specific instructions (that can speed up the calculation drastically) is checked at the initialization, but can also be deactivated completely.

In the software receiver implementation, the version 3.2.1 of FFTW is used as a dynamic library with double precision (*libfftw3-3.dll*). The FFT plans are created using the FFTW_MEASURE flag that tells FFTW to find an optimized plan by actually *computing* several FFTs and measuring their execution time. This operation can take some time but is executed only once at the beginning during the initialization phase when the receiver is not yet acquiring satellites.

The library with double-precision is compiled to take advantage of the SSE2 instructions (if available). In order to be able to use these instructions, the array of complex (or real) data passed to FFTW must be specially aligned in memory (typically 16 bit aligned). This must be done with the functions *fftw_malloc* (for allocating memory space) and *fftw_free* (for de-allocating the reserved memory space). These functions have exactly the same interface and behaviour as *malloc/free*, except that they ensure that the returned pointer has the correct alignment.

The code source (and also the compiled libraries for Windows) can be downloaded on the project homepage at http://www.fftw.org.

5.7.2 External header files

Table 5.9 gives an overview and a description of the used external header files and tries to list the functions that are using the provided functions.

File	Description	Used by
stdio.h	Structures, values, macros, and functions used by the standard I/O routines.	FFTW
stddef.h	Definitions and declarations for common constants, types, and variables.	FFTW
math.h	Mathematical library.	Several functions
windows.h	Master include for Windows applications.	Handles, events, ...
process.h	Function argument declarations for all process control related routines.	Threads
cyioctl.h	Cypress USB declarations.	Cypress controller
setupapi.h	Windows setup and device installer services.	Finding USB device

TABLE 5.9: External header files

5.8 Summary

This chapter presented the implementation of the new architecture developed in Chapter 4. This work has been one of the main activities during the development of the software receiver as the complete programming started from scratch.

Because it is – for the time being – not possible to work with the ideal software receiver architecture (see Section 3.5), some additional hardware parts were still needed. The first section described briefly these components and their goals, together with the needed programming and configuration.

Chapter 5 Implementation of new architecture and algorithms   119

The next section explained the USB interface, both on the front-end unit and on the host system side. The chosen configuration and details about the implementation were explained. This included a discussion about the different endpoints for sending and receiving data from and to the front-end unit as well as how the critical real-time data handling was implemented on the host side. For the latter, the methods for finding and initializing the device on the host system were first explained. Afterwards, the concept of using events and different threads for the data handling was presented and discussed in detail, together with a description of the implementation.

The third section covered the implementation of the baseband operations on the host system representing the core of the software receiver. The parameters and the algorithms for the different stages (acquisition, re-acquisition, and tracking) were explained and the final realization was discussed. The implementation of the new concept of using partial sums for the code and carrier removal process was explained in greater detail, starting with the computation and the storage of the partial sum vector. The characteristics and the implementation of the algorithms for generating and removing the code and the carrier in real-time were introduced and the advantages for the removal process were discussed. This section was concluded with a discussion about the realization of the code phase measurements and of the simplified tracking loops.

The last section showed the final software receiver prototype and introduced the concept of aiding and the corresponding implementation, together with an overview of external libraries that have been used for programming the solution.

Chapter 6

Setup, tests, and results

6.1 Introduction

This section describes the tests of the software receiver and discusses the obtained results.

To be able to measure and estimate the performance of the implemented software solution, the receiver is tested with two different signal sources:

1. Signal generated with a Spirent simulator (GSS8000);
2. Real signal coming from a fixed roof antenna.

The *simulated signal* allows testing the software receiver with known signals and well defined scenarios. The tests can be repeated several times with the same conditions giving a huge number of measurements that can be used to give a representative indication of the measured parameters. Nevertheless, it is important to notice that the simulated signals do not correspond always to the real signals as all errors sources can be eliminated and the signal can show a perfect shape.

The Spirent GSS8000 can simulate multiple satellite signals of a complete constellation and is used for testing and evaluating the performance of the whole software receiver, including the navigation solution. The simulated constellation can have the same characteristics and positions of the different satellites as the 'real-world' constellation, but can also represent a constellation that is never possible in the real world. Furthermore, the signal power levels can easily be changed allowing to test the sensitivity of the receiver. It is not only possible to simulate a static position (as seen with a fixed root antenna), but also a moving position of the receiver (with the conditions as in a car or an aircraft). With this device, the receiver can be evaluated as under real conditions (see limitation above) and the comportment of the different algorithms can be analyzed.

Real signals coming from a roof antenna allow testing and evaluating the software receiver under 'real' conditions, including all sources of error and multipath. But in this case, the position is always static.

The software receiver is tested for two aspects:

1. Accuracy and performance;
2. Requirements (processing time, CPU load, memory).

The *accuracy and performance tests* give an idea of the accuracy, the sensitivity, and the Time To First Fix (TTFF) of the receiver. These values allow comparing the performance of the implemented solution to commercial receivers found currently on the market and also show where optimizations can or must be applied. A more detailed description (like the parameters and the exact configuration) is given directly in the result section.

The *requirement tests* give an idea of the computational load and memory requirements of the implemented solution on different architectures. The time needed to post-process one minute of recorded data on different platforms is measured giving an indication of the brute processing load for the acquisition and tracking stages. It also shows if the receiver can handle the data processing in real-time. Additionally, the CPU load and memory requirements are also measured when the software receiver is working in real-time streaming mode (i.e., processing a real incoming data stream). These values are measured for the different stages (initialization, acquisition, and tracking) and on different platforms (Intel Atom N270, Intel Pentium 4, and Intel Core 2 Duo processor). A detailed description of the used hardware can be found in Appendix D.

6.2 Test setup

This section describes the test setups and the different tests that have been performed with the software receiver.

The test setup shown in Figure 6.1 is used for the simulated signals and the different components are described hereafter.

The Spirent GSS8000 simulator is used to generate the GPS signals for different satellites from a given scenario (although the GSS8000 could generate GPS and Galileo signals, only GPS is used for testing the software receiver). Connected to the Spirent is a power splitter from MiniCircuit (ZN2PD2-50-S+, [66]) allowing to feed the simulated signal simultaneously into the software receiver and into the u-blox 5 receiver that is used for aiding purposes. The software receiver front-end board is connected over USB to a Dell Precision 380 computer on which the software receiver is running. The u-blox 5 receiver is also connected over USB to this computer and is used only for aiding purposes during the initialization phase of the software receiver (getting initial time and ephemeris information). Once this information is obtained, the serial connection is no longer used.

A Dell Latitude D430 notebook is connected over Ethernet to the Spirent simulator and to the Dell Precision 380 computer. It acts as the master of the automatic test routines and has the following

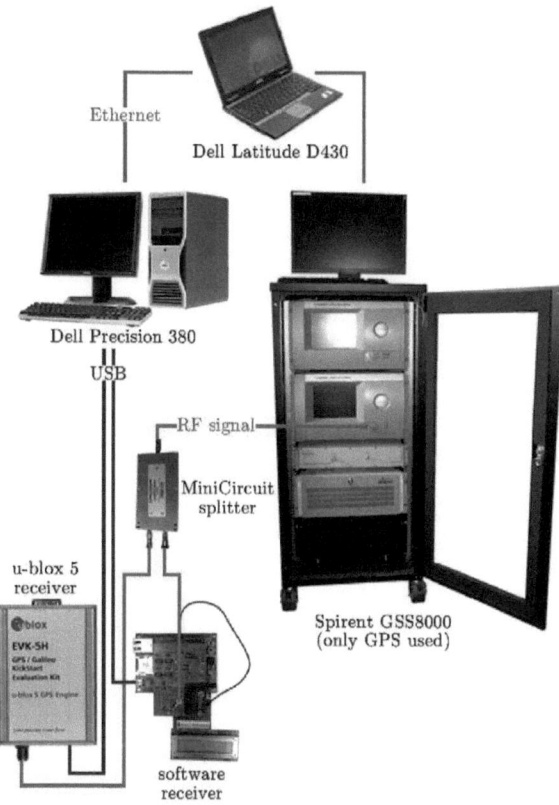

FIGURE 6.1: Test setup with simulator

tasks:

1. Loading the correct scenario on the Spirent simulator;
2. Running and stopping the scenarios on the Spirent simulator;
3. Running and stopping the software receiver on the Dell Precision 380;
4. Saving all the measurements.

If the tests are performed with real signals coming from the roof antenna, the Spirent simulator is replaced by the antenna. In this case, the starting and stopping of the scenarios becomes obsolete and the notebook serves mainly to save the measurements.

6.3 Description of the figures

This section describes the figures that are used for presenting the results. Figure 6.2 shows a template for the static tests, while Table 6.1 describes the different graphs.

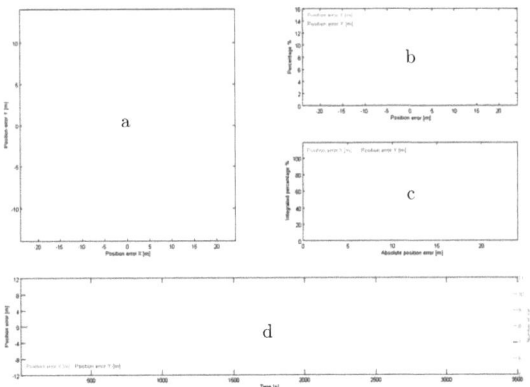

FIGURE 6.2: Template for presenting the results of static tests

Graph	Description
a	Deviation map with the position error for X versus Y, including the average and the standard deviation of the position error (for both X and Y). This is a superposition of all measurement points from all runs.
b	Histogram of the position error for X and Y, including the average errors. This is a superposition of all measurement points from all runs.
c	Integrated position error for X and Y, including the probability for 50% and 90%. This is a superposition of all measurement points from all runs.
d	Position error for X and Y versus time, including the number of satellites used by the navigation solution. This is the average of all measurement points from all runs.

TABLE 6.1: Description of the figures for static tests

The figures for the dynamic test consist in the same figures as for the static test, except that the deviation map (graph (a) in Figure 6.2) is replaced by a plot of the measured track coordinates X versus Y with the reference track superposed (in red).

The computation of the position error for the dynamic test is obtained by the following procedure:

1. The reference track was recorded on the Spirent simulator (with a resolution of 10 ms);
2. The first 100 position fixes of the software receiver were best matched to the positions of the reference track to achieve a time synchronization;

3. For each subsequent position fix, the absolute position error (for X and Y) was calculated with respect to the reference track.

6.4 Results

This section contains the results of the tests conducted with the software receiver. The detailed description of the different tests can be found directly in the subsection of the corresponding test.

6.4.1 Accuracy and performance tests (simulated signals)

Static position

During this test, the scenario was set to simulate a static position. After the specified duration, the simulation and the software receiver are stopped, the measurements saved and the test re-started, until the defined number of tests runs was reached. Table 6.2 gives the parameters used for the test with a static position.

Parameter	Value
Latitude	N 46° 54'
Longitude	E 69° 0'
Height	500.00 m
Start of simulation	23-Sep-2009 00:00:00
Duration of simulation	3600 seconds
Number of test runs	21

TABLE 6.2: Parameters for scenario with static position

The number of simulated satellites in view is given in Figure 6.3.

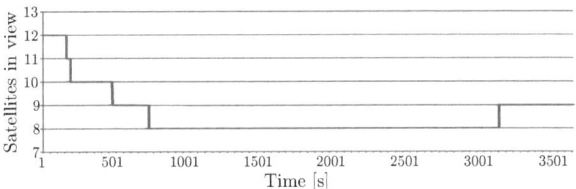

FIGURE 6.3: Static test: satellites in view

The simulated satellite constellation, the channel alignment, and the power levels are given in Figure 6.4. The nominal power level is set to -130 dBm which corresponds to the minimal signal strength on Earth. The final power level of each satellite depends on its position in the constellation.

Chapter 6 Setup, tests, and results 127

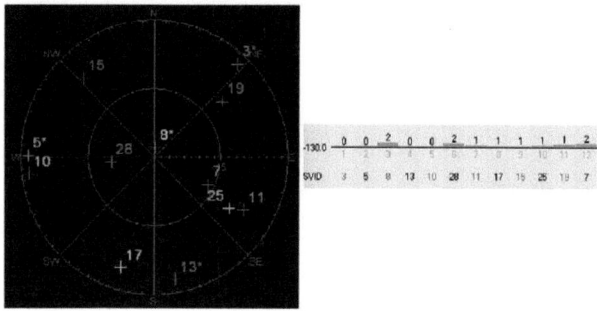

FIGURE 6.4: Static test: satellite constellation and power levels

The results of all 21 runs is given in Figure 6.5 while the results of the time-to-first-fix measurements are given in Table 6.3.

	Maximum [s]	Minimum [s]	Average [s]
Time to first fix	13.44	8.44	10.48

TABLE 6.3: Time-to-first-fix for scenario with static position

The following observations can be made:

1. The initial constellation contains 3 satellites with a very low elevation angle (satellites 3, 5, and 10). These satellites progressively disappear during the simulation and the changes can be clearly observed in the results (see next point).

2. These 3 satellites strongly influence the position and the latter systematically "jumps" a few meters when one of them disappears. This confirms the dependence of the initial position bias with respect to the constellation geometry.

3. The absolute position error (for X and Y) is kept within 20 meters for the entire 21 runs. This value was expected, as the navigation solution of the implemented software receiver was not tuned for high precision.

4. The position bias evolved over time in a quasi linear way, influenced by the geometry changes of the constellation. However, it can be assumed that the average of the bias over a longer observation period tends to zero. This hypothesis was confirmed with the real signal tests (see Section 6.4.2).

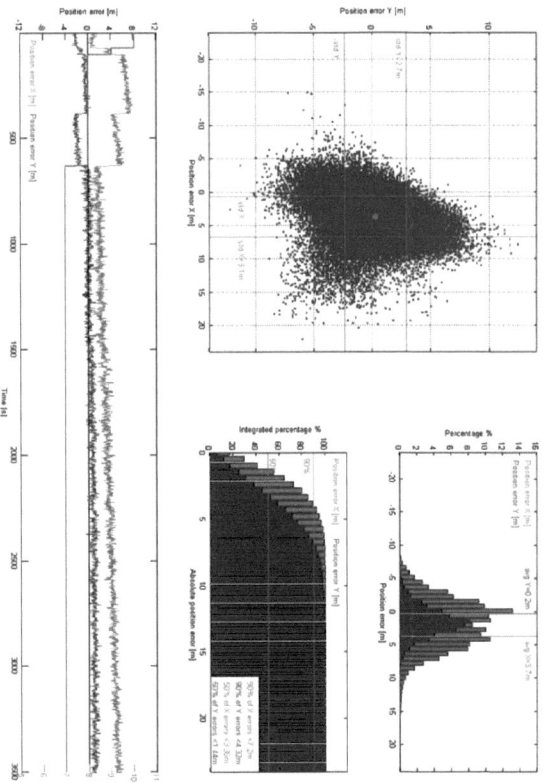

FIGURE 6.5: Static test: results

Dynamic position

During this test, the scenario was set to simulate a dynamic position. The duration of the simulation was 3600 seconds and a total of 10 test runs were performed. The racetrack has to properties shown in Figure 6.6.

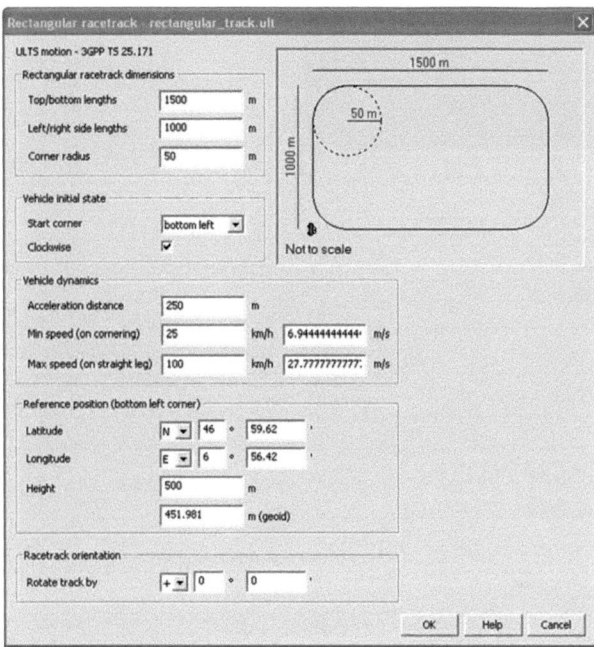

FIGURE 6.6: Dynamic test: racetrack properties

The number of simulated satellites in view is given in Figure 6.7, while Figure 6.8 shows the constellation, the channel alignment, and the power levels of the corresponding simulated satellites. During the whole run, a sufficient number of visible satellites is available.

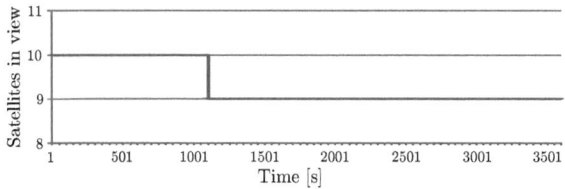

FIGURE 6.7: Dynamic test: satellites in view

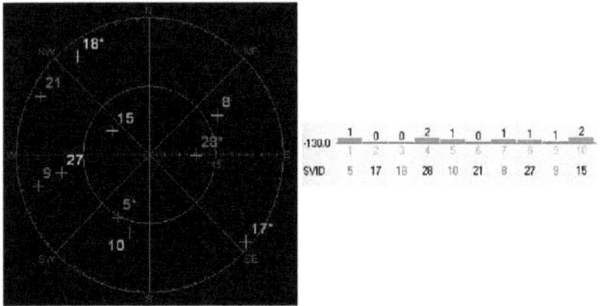

FIGURE 6.8: Dynamic test: satellite constellation and power levels

The result of a single run is given in Figure 6.9 while the results of the TTFF measurements are summarized in Table 6.4.

	Maximum [s]	Minimum [s]	Average [s]
Time to first fix	11.22	9.66	10.28

TABLE 6.4: Time-to-first-fix for scenario with dynamic position

The following observations can be made:

1. The position error is within 10 meters, which is comparable to the results obtained from the test with the static position.
2. The biggest errors are observed when the vehicle drives along the rounded corners of the rectangle. This translates into the "saw tooth" appearance of the time plot.
3. The time synchronization is achieved by best matching the first 100 position fixes of the software receiver with the positions of the reference track (with 10 ms resolution).
4. During the complete run, all available satellites have been tracked.

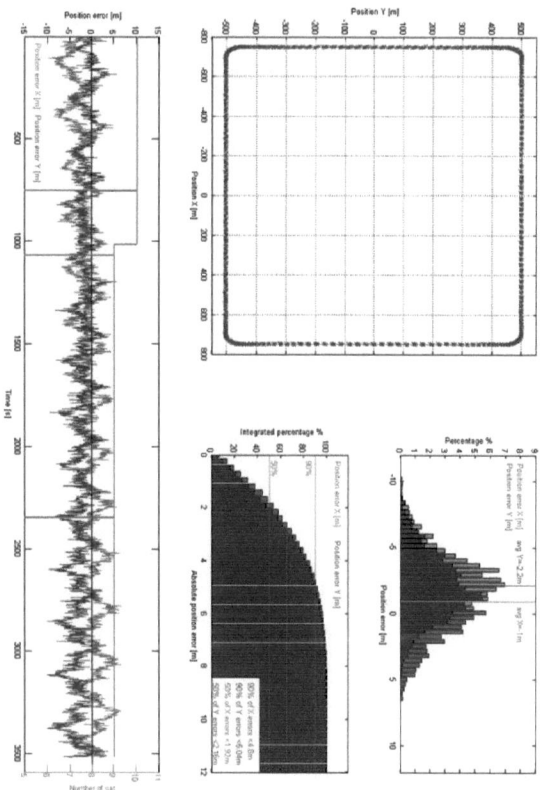

FIGURE 6.9: Dynamic test: results

6.4.2 Accuracy and performance tests (real signals)

During this test, the software receiver is connected to an Antcom 3GNSSA-XTR-1 (active L1/L2/L5-band) antenna that is installed on the roof of building of the Institute. Every measurement run is done for up to 8 hours with a total of 12 runs. Table 6.5 gives the parameters used for the test with a real signal.

Parameter	Value
Latitude	N 46.99385° (WGS-84 grid)
Longitude	E 6.9405° (WGS-84 grid)
Start of test	02-Nov-2009 13:44:35
Duration of test	up to 8 hours
Number of test runs	12

TABLE 6.5: Parameters for real signal test

As at least five satellites are required to perform a valid position fix, the analysis of the results is stopped as soon as less than 5 satellites are tracked by the software receiver. This happens normally after approximately 3 hours.

The superposition of the results of all 12 runs is given in Figure 6.10 while the results of the TTFF measurements for all 12 runs are given in Table 6.6.

	Maximum [s]	Minimum [s]	Average [s]
Time-to-first-fix	22.22	7.8	12.37

TABLE 6.6: Time-to-first-fix for real signal tests

The following observations can be made:

1. The position error distribution shows several linear behavior (dotted black line on position error graph) coming from the multipath components on the signal. Depending on the constellation, this error moves the position, but always more or less on a straight line. This effect can be seen more clearly if the results of only a single run is analyzed.
2. The average position error is quite small (< 5 meters) which comes again from the superposition of several constellations. But it shows also that the software receiver performs well with real signals.
3. The average position bias over all runs tends to zero (see the time plot of Figure 6.10).
4. After approximately 3 hours of run, the receiver lost all satellites, except of 5. For this reason, the analysis of the measurements is stopped.

Chapter 6 Setup, tests, and results 133

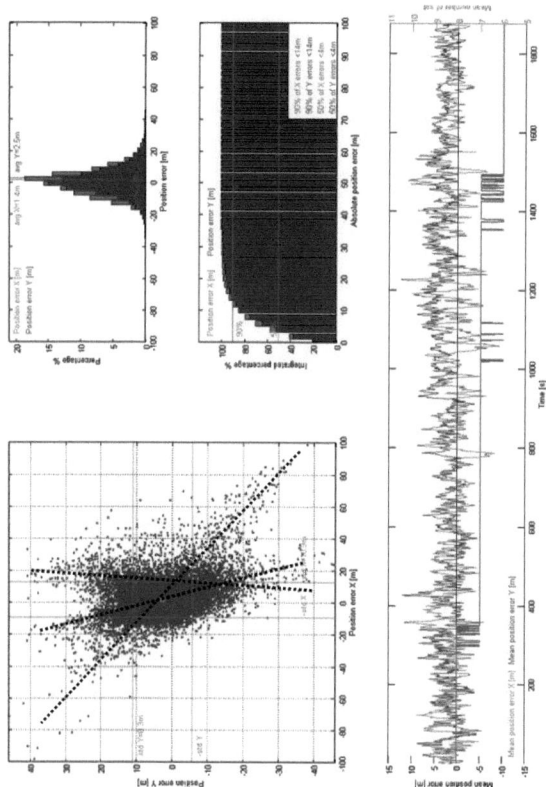

FIGURE 6.10: Real signal test: long time run

6.4.3 Requirement tests

Post-processing mode

The main goal of this test is to evaluate the complexity of the software receiver. This is done by measuring the computation time needed to post-process one minute of saved data from a file.

The complete data file ($\approx$ 650 MB) is first transferred from hard drive into the RAM to eliminate the read access delay caused by the disk controller. The software receiver then starts the processing of the data and the time spent for both the acquisition and the tracking stages is measured. This procedure is repeated with 1 up to 12 acquisition and tracking channels on different platforms with different processor types and architectures (see Appendix D for details about the used platforms).

The time needed in the acquisition stage (for the post-FFT and the FFT implementation) is shown in Figure 6.11 and Figure 6.12, while the time needed in the tracking stage is shown in Figure 6.13.

The following observations can be made:

1. With the Intel Core 2 Duo CPU and for a 12 channel configuration, the 60 seconds of data are post-processed in less than 8 seconds. This makes the execution more than 7 times faster than real-time.
2. Real-time performance can be achieved with this architecture even for the slowest CPU (Intel Atom N270) which processes all the data in less than 25 seconds. This confirms the extremely fast execution of the code and the efficiency of the implemented baseband algorithms (batch processing).
3. The Intel Atom CPU builds on an architecture that is comparable to the one from the Intel Pentium 4 processor. For a CPU frequency that is two times smaller (1.6 ↔ 3.2 GHz), the processing time on the Atom configuration is approximately multiplied by a factor of 2 with respect to the Pentium 4 execution time.
4. The AMD Athlon 2.21 GHz CPU is said to have roughly the same performance as an Intel Pentium 4 CPU clocked at 3.7 GHz. This statement can be verified from the figure.
5. The processing time shows an almost linear behavior with respect to the number of channels.
6. The re-acquisition stage (Figure 6.12) is extremely fast as it is only used to re-confirm the presence of the signal and the FFT is performed on a single frequency bin (found by the acquisition stage).

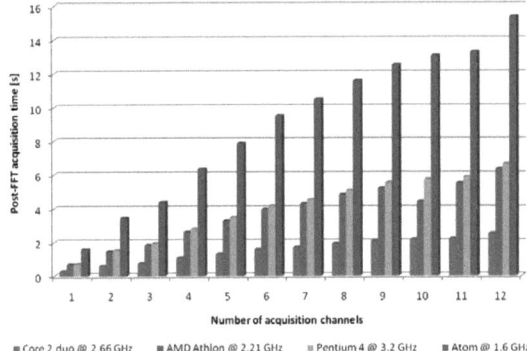

FIGURE 6.11: Requirement test: time of acquisition stage (post-FFT)

CPU load and memory requirements

The main goal of this test is to measure the CPU load and the memory requirements of the software receiver. For this reason, the code is launched on different platforms and the CPU load and the

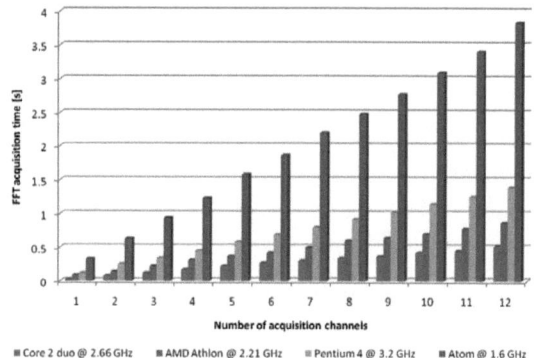

FIGURE 6.12: Requirement test: time of re-acquisition stage (FFT)

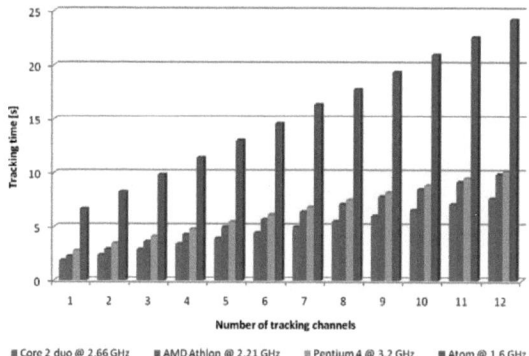

FIGURE 6.13: Requirement test: time of tracking stage

memory requirements are analyzed in detail. The software receiver is running with real data and the USB framework is activated.

The input signal is coming from a Spirent GSS8000 simulator and 10 channels are used in the software receiver. The following results are obtained for the different platforms: see Figure 6.14 – 6.17.

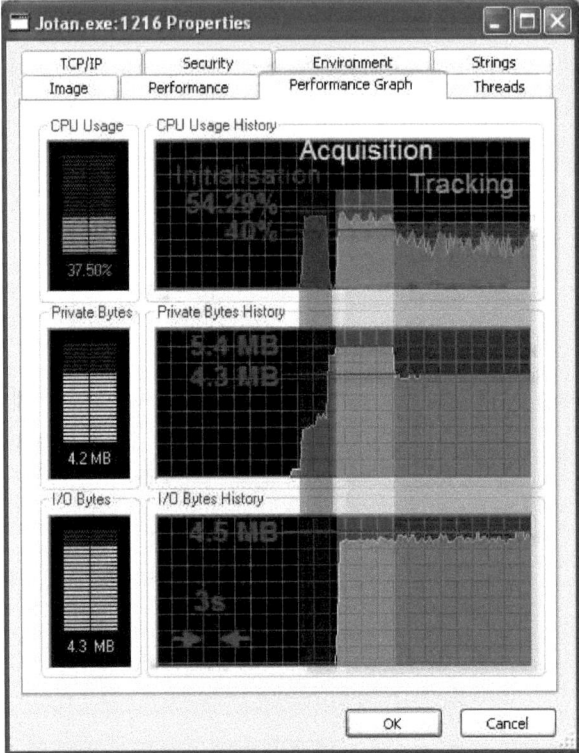

FIGURE 6.14: Requirement test: CPU load and memory requirement on an Intel Atom N270 platform

Table 6.7 summarizes the results.

The tool *ProcessExplorer* from Microsoft was used for the measurements [67]. A CPU load of 100% in the ProcessExplorer means that the software receiver fully loads all available cores.

The FFTW library in the acquisition phase generates the highest CPU load. As the FFTW library is not used with the multithreading option, the load is not distributed to the different CPUs and therefore, a maximum load of 50% is obtained.

6.5 Summary

This chapter described the different test performed with the software receiver and discussed the obtained results. It is important to keep in mind that the software receiver was developed during this

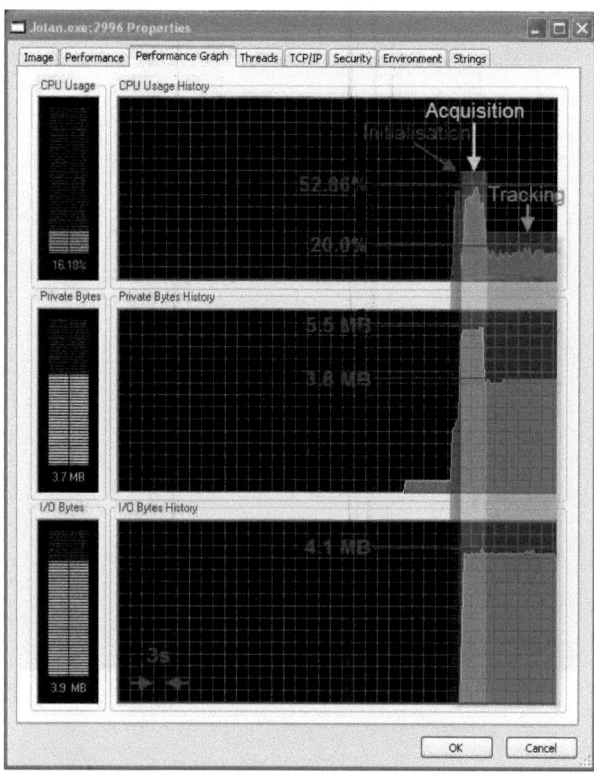

FIGURE 6.15: Requirement test: CPU load and memory requirement on Intel Mobile Core 2 Duo platform

	CPU load [%]		**Memory [MB]**	
	Acquisition	Tracking	Acquisition	Tracking
Atom N270	54.29	40	5.4	4.3
Mobile Core 2 Duo	52.86	20	5.5	3.8
Pentium 4	50	15.15	5.5	4.2
Core 2 Duo	50	11	5.8	4.3

TABLE 6.7: Requirements test: overview of results

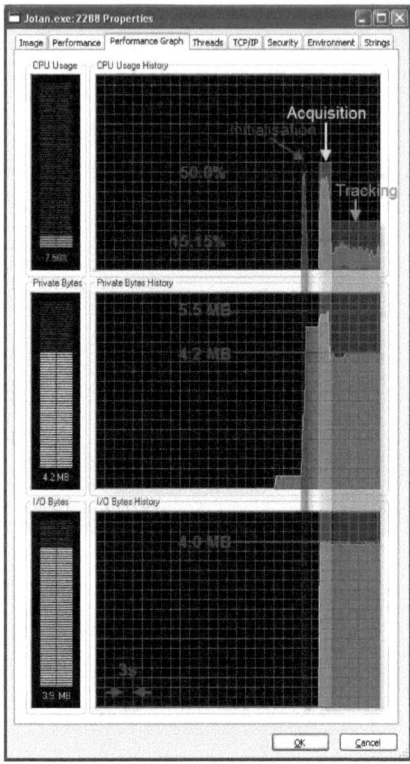

FIGURE 6.16: Requirement test: CPU load and memory requirement on Intel Pentium 4 platform

thesis with the goal of finding a solution of a real-time capable implementation on a general purpose microcontroller (thus, avoiding special instructions) and not of providing a receiver with short acquisition time or high accuracy.

The receiver was tested with simulated and real signals. This allowed giving a statistical interpretation of the obtained results as well as showing the behavior with real signals (including errors and multipath). Not only the accuracy and the performance (i.e., TTFF) were measured, but also the requirements in terms of CPU load and memory footprint.

The software receiver performed very well during the different tests. The accuracy for a static and a dynamic position with simulated signals was sufficient and even better than expected, given the fact that the receiver was not optimized for precision (neither the tracking loops nor the navigation solution). The accuracy of the position with real signals was decreased as some multipath components were present and no mitigation techniques have been implemented. The results of the TTFF

Chapter 6 Setup, tests, and results 139

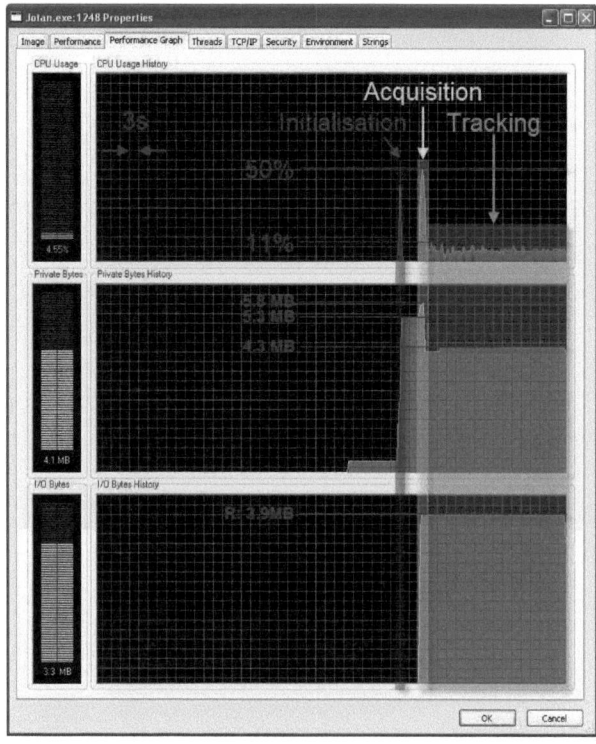

FIGURE 6.17: Requirement test: CPU load and memory requirement on Intel Core 2 Duo Platform

measurements showed no outstanding performance (for an aided receiver), but again, the goal of the implementation was not a fast acquisition time. The different stages (acquisition and tracking) have been verified successfully.

The more interesting and exciting results have been obtained for the requirement tests. The implementation showed its real potential and performance in the post-processing mode when one minute of raw data was processed in less than 25 seconds (with 12 channels) on a current microprocessor. This underlined the efficiency of the new batch-processing architecture (even without optimizations of the code or using multiple cores). The CPU load of the implemented solution was lower than expected (also compared to commercial products) and allowed running the solution even on a low-cost microprocessor. The memory footprint was really impressive as the complete solution did not occupied more than 4.5 MB of RAM. This was mainly due to the fact that no oversampled or pre-generated code or carrier replicas have to be stored in the memory and everything was calculated in real-time.

Chapter 7

Conclusion and Outlook

7.1 Conclusion

The overall aim of this work consisted in developing and implementing a real-time capable software receiver on a general purpose microprocessor. The term *general purpose microprocessor* implied that CPU specific functions (like SIMD operations and the distribution on multiple cores) had to be avoided as much as possible.

The main objectives were to:

1. Develop a new architecture and the associated algorithms optimized for the implementation on a general purpose microprocessor;
2. Implement the new architecture and algorithms;
3. Build a software receiver prototype;
4. Validate and test the prototype.

The development of the new architecture was introduced and explained in Chapter 4. The solution avoided CPU specific instruction and entirely employed integer operations. The resulting architecture is completely based on the work performed during the elaboration of this thesis. To be able to propose a new solution for a GNSS software receiver, a thorough understanding of the receiver functions was required. Therefore, Chapter 2 was written with the idea to discuss and explain the different elements and stages of a general GNSS receiver.

Chapter 3 was written to provide first a history and a general definition of the term *Software Receiver*, as this expression caused and still causes often confusion. Some of the main challenges of a software receiver were discussed and explained later because it was difficult at the beginning to estimate and identify them. The main part of the chapter was dedicated to the description of existing solutions and their implementation, together with an estimation of the needed computational load. The main contribution was to gather the different solutions and to establish the comparison of the computational load.

The final implementation of the proposed architecture of Chapter 4 was described in detail in Chapter 5. This contained all the aspects needed for a real-time capable software receiver, including the uninterrupted data transmission and the baseband processing using the partial sum architecture. The complete implementation was performed in the frame of this thesis and started from scratch.

Chapter 6 presented the obtained results and performance of the final implementation. The tests have been conducted with simulated and real signals, measuring the accuracy (for a static and a dynamic position), the computational load, and the memory requirements. The results have been shortly discussed.

The final performance and the requirements demonstrated that the proposed architecture and algorithms are well suited for an implementation of a real-time software receiver. The code was written directly based on the presented architecture and no special optimizations have been applied (e.g., by measuring the execution time for the different functions with a profiler and then optimizing the execution time). The aim of the work can therefore be considered as achieved and a software receiver was implemented on a general purpose microprocessor.

7.2 Outlook

From the above discussion and from the notes in Chapter 5, several ameliorations can be seen (the points are not ordered by priority):

- ○ Analysis of the code with a profiler and optimization of the most time-consuming operations.
- ○ Implementation of some functions or critical blocks in a low-level language to reduce the number of needed operations.
- ○ Decoupling of the different channels and handling of every channel individually to re-acquire lost and acquire new satellites during runtime.
- ○ Eliminating the introduced error by summing up 8 samples per partial sum (e.g., by using two partial sum vectors (one summing up 8 samples and one summing up 7 samples) and take the correct partial sum).
- ○ Decoding of the complete navigation message, including the detection of the data-bit transition to extract the ephemeris data.
- ○ Tuning of the tracking loops and discriminators to increase the receiver sensitivity.
- ○ Tuning of the acquisition stage to improve the acquisition time.
- ○ Exploiting all available system resources (e.g., multi-core system, optimized operations, ...) by testing the host system at startup to improve the execution time.

Finally, the concept of a software receiver will hopefully gain high interest during the next few years as the available resources in mobile devices will continuously increase and because the world of satellite navigation is in fast change. Adaptions to the new signals and modulations can be easily performed with a software receiver because "only" some lines of code have to be changed. However, the argument of offering a low-cost solution could be revoked in the next years as the current prices for hardware GNSS chips and modules are still dropping.

Appendix A

Carrier-to-Noise Density

This appendix explains briefly some of the RF quantities and unities commonly used in discussion of GPS signal power levels. The carrier-to-noise density ratio may not be familiar to those without experience in spread-spectrum systems. Definitions for *dBm*, *dBW*, *dB·Hz*, C/N_0 are given and related to the more familiar signal-to-noise ratio (*SNR*).

A.1 Power in decibels (dBm, dBW)

Power can be expressed in decibels by forming the ration of the power to a reference power level. Typical reference levels are one watt or one milliwatt. A power level in decibel-milliwatts can be computed from a power expressed in milliwatts as:

$$P_{dBm} = 10 \cdot \log\left(\frac{P_{mW}}{1_{mW}}\right) \tag{A.1}$$

Similarly, power can be expressed in decibel-watts as:

$$\begin{aligned} P_{dBW} &= 10 \cdot \log\left(\frac{P_W}{1_W}\right) = 10 \cdot \log\left(\frac{P_{mW}}{1000_{mW}}\right) \\ &= 10 \cdot \log\left(\frac{1}{1000} + \frac{P_{mW}}{1_{mW}}\right) = -30 + 10 \cdot \log\left(\frac{P_{mW}}{1_{mW}}\right) \\ &= P_{dBm} - 30 \end{aligned} \tag{A.2}$$

A.2 Signal-to-Noise Ratio (SNR)

The Signal-to-Noise Ratio (SNR) is defined as the signal power divided by the noise power, with this ratio usually expressed in decibels. For a signal power (S) and a noise power (N) defined in common units of power, such as watts or milliwatts, the SNR is:

$$\text{SNR} = 10 \cdot \log\left(\frac{S}{N}\right) \tag{A.3}$$

If the units of power are already in common decibel units, such as dBW or dBm, then

$$\text{SNR} = S_{dBm} - N_{dBm} = S_{dBW} - N_{dBW}. \tag{A.4}$$

A.3 Thermal Noise

Since noise sources like thermal noise generate power in proportion of the bandwidth of the system in question, a method of describing the power level independent of the bandwidth is desirable. Power spectral density is a measure of power in each unit of the bandwidth.

Thermal noise has a constant power density. The power of thermal noise generated is a function of the temperature and the noise bandwidth and it is independent of the central frequency of that bandwidth. The noise power spectral density for noise is $k \cdot T$, where k is the Boltzmann constant and T is the absolute temperature. Boltzmann's constant is the ratio of the energy in a molecule to its temperature, where the units are joules per degree Kelvin:

$$k = 1.38 \cdot 10^{-23} \, \frac{\text{J}}{\text{K}} \tag{A.5}$$

We recall the Kelvin temperature scale is zero at absolute zero, where the motion of the molecules stops. The Celsius temperature scale is shifted such that the zero-point is the freezing point of water, but the step size of a degree Celsius and a degree Kelvin is the same. Conversion between the two can be made from the relationship: 0 °C = 273.15 K.

Ambient thermal noise is typically calculated at 294 K, or 20.85 °C. This is the reference generally taken as the effective noise temperature on the earth:

$$N_T = k \cdot T = 1.38 \cdot 10^{-23} \, \frac{\text{J}}{\text{K}} \cdot 294 \text{ K} = 4.06 \cdot 10^{-21} \text{ J} \tag{A.6}$$

This is an expression of the noise power spectral density in joules, a unit of energy. Since a watt is a joule-per-second, the expression can also be written as watts-per-hertz. Ambient thermal noise power spectral density is then:

$$N_T = 4.06 \cdot 10^{-21} \, \frac{\text{W}}{\text{Hz}} = -204 \, \frac{\text{dBW}}{\text{Hz}} = -174 \, \frac{\text{dBm}}{\text{Hz}} \tag{A.7}$$

Appendix A Carrier-to-Noise Density

A.4 Carrier-to-Noise Density (C/N0)

The carrier-to-noise density is defined as the carrier power divided by the noise power spectral density. To calculate the carrier-to-noise density then for a GPS receiver operating at the thermal noise floor ($N_0 = N_T$), the carrier power is needed. The C/A code GPS signal specification [12] gives a nominal value for the carrier power received at the surface of the earth, specifying this to be at a power level of -160 dBW or above. Using this carrier power and the thermal noise floor, the carrier-to-noise density can be calculated as:

$$\frac{C}{N_0} = \frac{-160 \text{ dBW}}{-204 \text{ dBW/Hz}} = 44 \text{ dB} \cdot \text{Hz} \tag{A.8}$$

Now, this value can be converted to a signal-to-noise ratio for the same conditions, assuming a C/A code receiver bandwidth of 2 MHz. To convert the carrier-to-noise density to the signal-to-noise ratio, the bandwidth needs to be divided:

$$\text{SNR} = \frac{-160 \text{ dBW}}{2 \cdot 10^6 \text{ Hz} \cdot \left(-204 \frac{\text{dBW}}{\text{Hz}}\right)} = \frac{-160 \text{ dBW}}{-141 \text{ dBW}} = -19 \text{ dB} \tag{A.9}$$

This means that the signal is 19 dB below the noise power coming into the receiver. It is only through de-spreading the C/A code that the signal can be detected at all. After the signal is de-spread, it is filtered to a narrower bandwidth. The decrease in bandwidth eliminates most of the noise which is spread over the entire bandwidth, but leaves the signal. This is the mechanism by which a *processing gain* is achieved in a spread-spectrum receiver. A more complete description of the process gain can be found in [68].

Appendix B

Phase-Locked Loops

This appendix explains the theory of the PLL, starting with the description of the basic PLL, followed by the explication of a first and second order PLL and concludes with the transformation from a continuous to a discrete systems. More details can be found in [11].

B.1 Basic Phase-Locked Loop

The main purpose of a PLL is to adjust the frequency of a local oscillator to match the frequency of the incoming signal. The structure of a basic PLL is shown in Figure B.1, with (a) the time domain and (b) the s-domain representation, which is obtained from the Laplace transform (as explained later in this section).

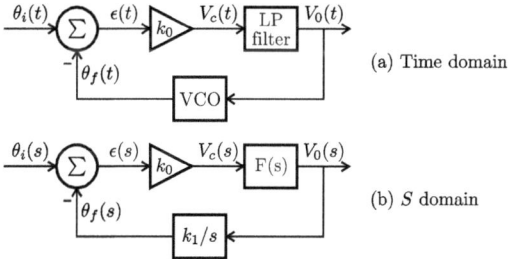

FIGURE B.1: Basic phase-locked loop (time- and s-domain)

The input signal is $\theta_i(t)$ and the output from the Voltage Controlled Oscillator (VCO) is $\theta_f(t)$. The phase comparator $\sum$ measures the phase difference of these two signals. The amplifier k_0 represents the gain of the phase comparator and the low-pass filter limits the noise in the loop. The input voltage

V_0 to the VCO controls its output frequency, which can be expressed as given in Equation B.1.

$$\omega_2(t) = \omega_0 + k_1 \cdot u(t) \cdot V_0 \tag{B.1}$$

where ω_0 is the center angular frequency of the VCO, k_1 is the gain of the VCO, and $u(t)$ is a unit step function, which is defined in Equation B.2.

$$u(t) = \begin{cases} 0 & \text{for } t < 0 \\ 1 & \text{for } t > 0 \end{cases} \tag{B.2}$$

The phase angle can be obtained by integrating Equation B.1 and gives the results as shown in Equation B.3.

$$\begin{aligned}\int_0^t \omega_2(t) dt &= \omega_0 \cdot t + \theta_f(t) \\ &= \omega_0 \cdot t + \int_0^t k_1 \cdot u(t) \cdot V_0\, dt\end{aligned} \tag{B.3}$$

$$\text{where} \quad \theta_f(t) = \int_0^t k_1 \cdot u(t) \cdot V_0\, dt$$

The Laplace transform of $\theta_f(t)$ is given in Equation B.4.

$$\theta_f(s) = V_0(s) \cdot \frac{k_1}{s} \tag{B.4}$$

From Figure B.1 the Equation B.5 can be written.

$$\begin{aligned}V_c(s) &= k_0 \cdot \epsilon(s) = k_0 \cdot [\theta_i(s) - \theta_f(s)] \\ V_0(s) &= V_c(s) \cdot F(s) \\ \theta_f(s) &= V_0(s) \cdot \frac{k_1}{s}\end{aligned} \tag{B.5}$$

From Equation B.5, the error function $\epsilon(s)$ and the input signal $\theta_i(s)$ can be obtained, as given in Equation B.6.

$$\begin{aligned}\epsilon(s) = \theta_i(s) - \theta_f(s) &= \frac{V_c(s)}{k_0} = \frac{V_0(s)}{k_0 \cdot F(s)} \\ &= \frac{s \cdot \theta_f(s)}{k_0 \cdot k_1 \cdot F(s)} \\ \theta_i(s) &= \theta_f(s) \cdot \left(1 + \frac{s}{k_0 \cdot k_1 \cdot F(s)}\right)\end{aligned} \tag{B.6}$$

Appendix B Phase-Locked Loops

The transfer function $H(s)$ of the loop is defined as given in Equation B.7.

$$\begin{aligned} H(s) &\equiv \frac{\theta_f(s)}{\theta_i(s)} \\ &= \frac{k_0 \cdot k_1 \cdot F(s)}{s + k_0 \cdot k_1 \cdot F(s)} \end{aligned} \qquad (B.7)$$

The error transfer function $H_e(s)$ is defined as given in Equation B.8.

$$\begin{aligned} H_e(s) &\equiv \frac{\epsilon(s)}{\theta_i(s)} \\ &= \frac{\theta_i(s) - \theta_f(s)}{\theta_i(s)} = 1 - H(s) \\ &= \frac{s}{s + k_0 \cdot k_1 \cdot F(s)} \end{aligned} \qquad (B.8)$$

The equivalent noise bandwidth is finally given in Equation B.9.

$$B_n = \int_0^\infty |H(j \cdot \omega)|^2 \, df \qquad (B.9)$$

where ω is the angular frequency that is related to the frequency as given in Equation B.10.

$$\omega = 2 \cdot \pi \cdot f \qquad (B.10)$$

In order to study the properties of the PLL, two different types of input signals are generally used. The first type is a step function as given in Equation B.11.

$$\begin{aligned} \theta_i(t) &= u(t) \\ \text{or} \quad \theta_i(s) &= \frac{1}{s} \end{aligned} \qquad (B.11)$$

The second type is a frequency modulated signal as given in Equation B.12.

$$\theta_i(t) = \Delta\omega \cdot t \quad \text{or} \quad \theta_i(s) = \frac{\Delta\omega}{s^2} \qquad (B.12)$$

These two types of signals will be used for the discussion in the next two sections.

B.2 First order Phase-Locked Loop

The definition of a first order PLL implies that the denominator of the transfer function $H(s)$ is a first order function of s. Equation B.13 gives the filter function for this kind of PLL.

$$F(s) = 1 \tag{B.13}$$

This is the simplest form of a PLL. For a unit step function input, the corresponding transfer function from Equation B.7 is given in Equation B.14.

$$H(s) = \frac{k_0 \cdot k_1}{s + k_0 \cdot k_1} \tag{B.14}$$

The denominator of $H(s)$ is a first order of s.

The noise bandwidth can be found as given in Equation B.15.

$$\begin{aligned} B_n &= \int_0^\infty \frac{(k_0 \cdot k_1)^2}{\omega^2 + (k_0 \cdot k_1)^2} \, df \\ &= \frac{(k_0 \cdot k_1)^2}{2 \cdot \pi} \cdot \int_0^\infty \frac{1}{\omega^2 + (k_0 \cdot k_1)^2} \, df \\ &= \frac{(k_0 \cdot k_1)^2}{2 \cdot \pi \cdot k_0 \cdot k_1} \cdot \tan^{-1}\left(\frac{\omega}{k_0 \cdot k_1}\right) \Big|_0^\infty \\ &= \frac{k_0 \cdot k_1}{4} \end{aligned} \tag{B.15}$$

With the input signal $\theta_i(s) = 1/s$, the error function from Equation B.8 can be transformed into Equation B.16.

$$\begin{aligned} \epsilon(s) &= \theta_i(s) \cdot H_e(s) \\ &= \frac{1}{s + k_0 \cdot k_1} \end{aligned} \tag{B.16}$$

The steady-state error can be found from the final value theorem of the Laplace transform, which can be stated as given in Equation B.17.

$$\lim_{t \to \infty} y(t) = \lim_{s \to 0} s \cdot Y(s) \tag{B.17}$$

Using Equation B.17, the final value of $\epsilon(t)$ from Equation B.16 can be written as given in Equation B.18.

$$\begin{aligned} \lim_{t \to \infty} \epsilon(t) &= \lim_{s \to 0} s \cdot \epsilon(s) \\ &= \lim_{s \to 0} \frac{s}{s + k_0 \cdot k_1} \\ &= 0 \end{aligned} \tag{B.18}$$

Appendix B Phase-Locked Loops

With the input function $\theta_i(s) = \Delta\omega/s^2$, the same calculations can be applied and the error function is given in Equation B.19.

$$\epsilon(s) = \theta_i(s) \cdot H_e(s)$$
$$= \frac{\Delta\omega}{s} \cdot \frac{1}{s + k_0 \cdot k_1} \tag{B.19}$$

The steady-state error for the frequency modulated signal input is given in Equation B.20.

$$\lim_{t \to \infty} \epsilon(t) = \lim_{s \to 0} s \cdot \epsilon(s)$$
$$= \lim_{s \to 0} \frac{\Delta\omega}{s + k_0 \cdot k_1}$$
$$= \frac{\Delta\omega}{k_0 \cdot k_1} \tag{B.20}$$

It is important to notice that this steady-state error is not equal to zero. A large value of $k_0 \cdot k_1$ can make this error small. However, from Equation B.15, the 3 dB bandwidth occurs at $s = k_0 \cdot k_1$. Thus, a small final value of $\epsilon(t)$ also means a large bandwidth, which contains more noise.

B.3 Second order Phase-Locked Loop

A second order PLL means that the denominator of the transfer function $H(s)$ is a second order function of s. One of the filters to make such a second order phase-locked loop is given in Equation B.21.

$$F(s) = \frac{s \cdot \tau_2 + 1}{s \cdot \tau_1} \tag{B.21}$$

Substituting Equation B.21 into Equation B.7 gives the transfer function as given in Equation B.22.

$$H(s) = \frac{\frac{k_0 \cdot k_1 \cdot \tau_2 \cdot s}{\tau_1} + \frac{k_0 \cdot k_1}{\tau_1}}{s^2 + \frac{k_0 \cdot k_1 \cdot \tau_2 \cdot s}{\tau_1} + \frac{k_0 \cdot k_1}{\tau_1}}$$
$$\equiv \frac{2 \cdot \zeta \cdot \omega_n \cdot s + \omega^2}{s^2 + 2 \cdot \zeta \cdot \omega_n \cdot s + \omega_n^2} \tag{B.22}$$

where ω_n is the natural frequency as expressed in Equation B.23.

$$\omega_n = \sqrt{\frac{k_0 \cdot k_1}{\tau_1}} \tag{B.23}$$

and ζ the damping factor as expressed in Equation B.24.

$$2 \cdot \zeta \cdot \omega_n = \frac{k_0 \cdot k_1 \cdot \tau_2}{\tau_1}$$
$$\zeta = \frac{\omega_n \cdot \tau_2}{2}$$
(B.24)

The damping ratio controls how fast the filter reaches its settle point and how much overshoot the filter can have; a smaller settling time results in a larger overshoot. Therefore, the choice of damping ratio is a compromise between overshoot and settling time. A good value for the damping ratio is $\zeta = 0.707$ resulting in a filter that converges reasonably fast and does not make a high overshoot.

The noise bandwidth can be found as given in Equation B.25 [69].

$$\begin{aligned} B_n &= \int_0^\infty |H(\omega)|^2 \, df \\ &= \frac{\omega_n}{2 \cdot \pi} \cdot \int_0^\infty \frac{1 + \left(2 \cdot \zeta \cdot \frac{\omega}{\omega_n}\right)^2}{\left[1 - \left(\frac{\omega}{\omega_n}\right)^2\right]^2 + \left(2 \cdot \zeta \cdot \frac{\omega}{\omega_2}\right)^2} \, d\omega \\ &= \frac{\omega_n}{2 \cdot \pi} \cdot \int_0^\infty \frac{1 + 4 \cdot \zeta^2 \cdot \left(\frac{\omega}{\omega_n}\right)^2}{\left(\frac{\omega}{\omega_n}\right)^4 + 2 \cdot (2 \cdot \zeta^2 - 1) \cdot \left(\frac{\omega}{\omega_n}\right)^2 + 1} \, d\omega \\ &= \frac{\omega_n}{2} \cdot \left(\zeta + \frac{1}{4 \cdot \zeta}\right) \end{aligned}$$
(B.25)

The mathematical development of this integration can be found in [11].

The error transfer function can be obtained from Equation B.8 and results in Equation B.26.

$$\begin{aligned} H_e(s) &= 1 - H(s) \\ &= \frac{s^2}{s^2 + 2 \cdot \zeta \cdot \omega_n \cdot s + \omega_n^2} \end{aligned}$$
(B.26)

Equation B.27 gives the error function for an input signal of $\theta_i(s) = 1/s$.

$$\epsilon(s) = \frac{s}{s^2 + 2 \cdot \zeta \cdot \omega_n \cdot s + \omega_n^2}$$
(B.27)

and the steady-state error is given in Equation B.28.

$$\lim_{t \to \infty} \epsilon(t) = \lim_{s \to 0} s \cdot \epsilon(s)$$
$$= 0$$
(B.28)

Appendix B Phase-Locked Loops

Equation B.29 gives the error function when the input is $\theta_i(s) = 1/s^2$.

$$\epsilon(s) = \frac{1}{s^2 + 2 \cdot \zeta \cdot \omega_n \cdot s + \omega_n^2} \tag{B.29}$$

and the steady-state error is given in Equation B.30.

$$\lim_{t \to \infty} \epsilon(t) = \lim_{s \to 0} s \cdot \epsilon(s) \tag{B.30}$$
$$= 0$$

In contrast to the first order loop, the steady-state error is zero for the frequency-modulated signal. This means the second order loop is able to track a modulated signal and returns the phase comparator characteristic to the null point. For this reason, the conventional PLL in a GPS receiver is usually a second order one.

A third order PLL can also be implemented but will not be discussed in this document. For details about the digital implementation, please refer to [10].

B.4 Transform from continuous to discrete systems

In the previous sections, the discussion was based on continuous systems. As a GNSS receiver works with digitized data, the continuous system has to be changed into a discrete system. The transfer from the s-domain into the discrete z-domain is done through the bilinear transform as given in Equation B.31.

$$s = \frac{2}{t_s} \cdot \frac{1 - z^{-1}}{1 + z^{-1}} \tag{B.31}$$

where t_s is the sampling interval.

Equation B.31 can be substituted into Equation B.21 and the filter function in the z-domain is given in Equation B.32.

$$F(z) = C_1 + \frac{C_2}{1 - z^{-1}}$$
$$= \frac{(C_1 + C_2) - C_1 \cdot z^{-1}}{1 - z^{-1}} \tag{B.32}$$

$$\text{where} \quad C_1 = \frac{2 \cdot \tau_2 - t_2}{2 \cdot \tau_1}$$
$$C_2 = \frac{t_s}{\tau_1}$$

This discrete loop filter is depicted in Figure B.2.

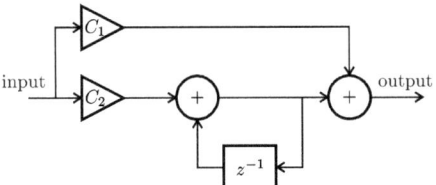

FIGURE B.2: Discrete loop filter

The VCO in the PLL is replaced by a direct digital frequency synthesizer and its transfer function $N(z)$ can be used to replace the result in Equation B.5 as given in Equation B.33.

$$N(z) = \frac{\theta_f(z)}{V_0(z)} \\ \equiv \frac{k_1 \cdot z^{-1}}{1 - z^{-1}} \tag{B.33}$$

Using the same approach as in Equation B.8, the transfer function $H(z)$ can be written as given in Equation B.34.

$$H(z) = \frac{\theta_f(z)}{\theta_i(z)} \\ = \frac{k_0 \cdot F(z) \cdot N(z)}{1 + k_0 \cdot F(z) \cdot N(z)} \tag{B.34}$$

Substituting the results from Equation B.32 and Equation B.33 into Equation B.34, the result can be written as given in Equation B.35.

$$H(z) = \frac{k_0 \cdot k_1 \cdot (C_1 + C_2) \cdot z^{-1} - k_0 \cdot k_1 \cdot C_1 \cdot z^{-2}}{1 + [k_0 \cdot k_1 \cdot (C_1 + C_2) - 2] \cdot z^{-1} + (1 - k_0 \cdot k_1 \cdot C_1) \cdot z^{-2}} \tag{B.35}$$

By applying bilinear transform (Equation B.31) to Equation B.22, the result can be written as given in Equation B.36.

$$H(z) = \frac{(4 \cdot \Xi + \Omega^2) + 2 \cdot \Omega^2 \cdot z^{-1} + (\Omega^2 - 4 \cdot \Xi) \cdot z^{-2}}{(4 + 4 \cdot \Xi + \Omega^2) + (2 \cdot \Omega^2 - 8) \cdot z^{-1} + (4 - 4 \cdot \Xi + \Omega^2) \cdot z^{-2}} \tag{B.36}$$

where $\Omega = \omega_n \cdot t_s$
$\Xi = \zeta \cdot \Omega = \zeta \cdot \omega_n \cdot t_s$

Appendix B Phase-Locked Loops

By equating the denominator polynomials in the above two equations, C_1 and C_2 can be found as given in Equation B.37.

$$C_1 = \frac{1}{k_0 \cdot k_1} \cdot \frac{8 \cdot \zeta \cdot \omega_n \cdot t_s}{4 + 4 \cdot \zeta \cdot \omega_n \cdot t_s + (\omega_n \cdot t_s)^2}$$
$$C_2 = \frac{1}{k_0 \cdot k_1} \cdot \frac{4 \cdot (\omega_n \cdot t_s)^2}{4 + 4 \cdot \zeta \cdot \omega_n \cdot t_s + (\omega_n \cdot t_s)^2} \quad \text{(B.37)}$$

Appendix C

Signal description baseband architectures

This appendix contains the signal description and the mathematical development for the carrier removal process in the real and in the complex baseband architecture, as described in Section 3.7.1. Most of the calculations are based on the relations given in Equation C.1.

$$\begin{aligned}
\sin(\alpha) \cdot \sin(\beta) &= \frac{1}{2} \cdot [\cos(\alpha - \beta) - \cos(\alpha + \beta)] \\
\cos(\alpha) \cdot \cos(\beta) &= \frac{1}{2} \cdot [\cos(\alpha - \beta) + \cos(\alpha + \beta)] \\
\sin(\alpha) \cdot \cos(\beta) &= \frac{1}{2} \cdot [\sin(\alpha - \beta) + \sin(\alpha + \beta)] \\
\sin(-\alpha) &= -\sin(\alpha)
\end{aligned} \quad (C.1)$$

To simplify the notation of the following equations, the variables as defined in Table C.1 are used.

Variable	Description
A_{CA}	Amplitude of the signal $A_{CA} = \sqrt{2 \cdot P_{CA}} \cdot (C(t) \otimes D(t))$
P_{CA}	Power of the C/A signal
$C(t)$	C/A code sequence
$D(t)$	Navigation data sequence
$\Delta\omega$	Difference between IF and carrier frequency $\Delta\omega = \omega_{if} - \omega_{car}$
$\Sigma\omega$	Sum of IF and carrier frequency $\Sigma\omega = \omega_{if} + \omega_{car}$
$\Delta\phi$	Difference between IF and carrier phase $\Delta\phi = \phi_{if} - \phi_{car}$
$\Sigma\phi$	Sum of IF and carrier phase $\Sigma\phi = \phi_{if} + \phi_{car}$

TABLE C.1: Variables for carrier and Doppler removal in real and complex architecture

C.1 Real baseband architecture

The real baseband architecture is shown in Figure 3.4, together with the notation of the different signal branches. By analyzing the architecture, the signal description in Equation C.2 can be found.

$$I = A_{CA} \cdot \cos(\omega_{if} \cdot t + \phi_{if})$$

$$\begin{aligned}
II &= I \cdot \cos(\omega_{car} \cdot t + \phi_{car}) \\
&= A_{CA} \cdot \cos(\omega_{if} \cdot t + \phi_{if}) \cdot \cos(\omega_{car} \cdot t + \phi_{car}) \\
&= \frac{A_{CA}}{2} \cdot \Big[\cos((\omega_{if} - \omega_{car}) \cdot t + \phi_{if} - \phi_{car}) + \\
&\qquad \cos((\omega_{if} + \omega_{car}) \cdot t + \phi_{if} + \phi_{car})\Big] \\
&= \frac{A_{CA}}{2} \cdot \Big[\cos(\Delta\omega \cdot t + \Delta\phi) + \cos(\Sigma\omega \cdot t + \Sigma\phi)\Big]
\end{aligned} \quad (C.2)$$

$$\begin{aligned}
IQ &= I \cdot \sin(\omega_{car} \cdot t + \phi_{car}) \\
&= A_{CA} \cdot \cos(\omega_{if} \cdot t + \phi_{if}) \cdot \sin(\omega_{car} \cdot t + \phi_{car}) \\
&= \frac{A_{CA}}{2} \cdot \Big[\sin((\omega_{car} - \omega_{if}) \cdot t + \phi_{car} - \phi_{if}) + \\
&\qquad \sin((\omega_{if} + \omega_{car}) \cdot t + \phi_{if} + \phi_{car})\Big] \\
&= -\frac{A_{CA}}{2} \cdot \Big[\sin(\Delta\omega \cdot t + \Delta\phi) + \sin(\Sigma\omega \cdot t + \Sigma\phi)\Big]
\end{aligned}$$

The amplitude in the two different branches (and therefore in the accumulators) is reduced by a factor 1/2. This means that each of the accumulator receives only half of the inital amplitude of the incoming signal.

C.2 Complex baseband architecture

The complex baseband architecture is shown in Figure 3.5, together with the notation of the different signal branches. By analyzing the architecture, the signal description in Equations C.3 – C.4 can be found.

$$I = A_{CA} \cdot \cos(\omega_{if} \cdot t + \phi_{if})$$

$$\begin{aligned}
II &= I \cdot \cos(\omega_{car} \cdot t + \phi_{car}) \\
&= A_{CA} \cdot \cos(\omega_{if} \cdot t + \phi_{if}) \cdot \cos(\omega_{car} \cdot t + \phi_{car}) \\
&= \frac{A_{CA}}{2} \cdot \Big[\cos\big((\omega_{if} - \omega_{car}) \cdot t + \phi_{if} - \phi_{car}\big) + \\
&\qquad\qquad \cos\big((\omega_{if} + \omega_{car}) \cdot t + \phi_{if} + \phi_{car}\big)\Big] \\
&= \frac{A_{CA}}{2} \cdot \Big[\cos\big(\Delta\omega \cdot t + \Delta\phi\big) + \cos(\Sigma\omega \cdot t + \Sigma\phi)\Big]
\end{aligned} \qquad (C.3)$$

$$\begin{aligned}
IQ &= I \cdot \sin(\omega_{car} \cdot t + \phi_{car}) \\
&= A_{CA} \cdot \cos(\omega_{if} \cdot t + \phi_{if}) \cdot \sin(\omega_{car} \cdot t + \phi_{car}) \\
&= \frac{A_{CA}}{2} \cdot \Big[\sin\big((\omega_{car} - \omega_{if}) \cdot t + \phi_{car} - \phi_{if}\big) + \\
&\qquad\qquad \sin\big((\omega_{if} + \omega_{car}) \cdot t + \phi_{if} + \phi_{car}\big)\Big] \\
&= \frac{A}{2_{CA}} \cdot \Big[-\sin(\Delta\omega \cdot t + \Delta\phi) + \sin(\Sigma\omega \cdot t + \Sigma\phi)\Big]
\end{aligned}$$

Appendix C Signal description baseband architectures

$$Q = A_{CA} \cdot \sin(\omega_{if} \cdot t + \phi_{if})$$

$$\begin{aligned}
QI &= I \cdot \cos(\omega_{car} \cdot t + \phi_{car}) \\
&= A_{CA} \cdot \sin(\omega_{if} \cdot t + \phi_{if}) \cdot \cos(\omega_{car} \cdot t + \phi_{car}) \\
&= \frac{A_{CA}}{2} \cdot \Big[\sin\big((\omega_{if} - \omega_{car}) \cdot t + \phi_{if} - \phi_{car}\big) + \\
&\quad\quad \sin\big((\omega_{if} + \omega_{car}) \cdot t + \phi_{if} + \phi_{car}\big) \Big] \\
&= \frac{A_{CA}}{2} \cdot \Big[\sin(\Delta\omega \cdot t + \Delta\phi) + \sin(\Sigma\omega \cdot t + \Sigma\phi) \Big]
\end{aligned} \quad (C.4)$$

$$\begin{aligned}
QQ &= I \cdot \sin(\omega_{car} \cdot t + \phi_{car}) \\
&= A_{CA} \cdot \sin(\omega_{if} \cdot t + \phi_{if}) \cdot \sin(\omega_{car} \cdot t + \phi_{car}) \\
&= \frac{A_{CA}}{2} \cdot \Big[\cos\big((\omega_{if} - \omega_{car}) \cdot t + \phi_{if} - \phi_{car}\big) - \\
&\quad\quad \cos\big((\omega_{if} + \omega_{car}) \cdot t + \phi_{if} + \phi_{car}\big) \Big] \\
&= \frac{A_{CA}}{2} \cdot \Big[\cos(\Delta\omega \cdot t + \Delta\phi) - \cos(\Sigma\omega \cdot t + \Sigma\phi) \Big]
\end{aligned}$$

After combining the real and imaginary part, the term given in Equation C.5 is obtained.

$$\begin{aligned}
II + QQ &= A_{CA} \cdot \cos(\Delta\omega \cdot t + \Delta\phi) \\
QI - IQ &= A_{CA} \cdot \sin(\Delta\omega \cdot t + \Delta\phi)
\end{aligned} \quad (C.5)$$

In comparison with Equation C.2, the whole amplitude A of the original incoming signal is used in the accumulators and the high frequency components (with the terms $\Sigma\omega$ and $\Sigma\phi$) are automatically canceled.

Appendix D

Platform specifications

This appendix describes the specifications for the different platforms that have been used for testing and evaluating the software receiver.

D.1 Asus EeePC 1000H (Notebook)

Component	Description
CPU type	Intel Atom N270, 1.6 GHz
CPU original clock speed	1600 MHz
CPU L1 code cache	32 kB
CPU L2 cache	512 kB (on-Die, ATC, full-speed)
System memory	2048 MB
Disk drive	Seagate ST9160827AS (150 GB, IDE)

D.2 Dell Latitude D430 (Notebook)

Component	Description
CPU type	Mobile Intel Core 2 Duo (Merom-2M)
CPU original clock speed	1200 MHz
CPU L1 code cache	32 kB per core
CPU L1 data cache	32 kB per core
CPU L2 cache	2 MB (on-Die, ASC, full-speed)
System memory	2048 MB
System memory speed	533 MHz
Disk drive	Samsung PZA032 SSD (32 GB, IDE)

D.3 Dell Precision 380 (Single core desktop)

Component	Description
CPU type	Intel Pentium 4 640 (Prescott-2M)
CPU original clock speed	3200 MHz
CPU L1 trace cache	12 kB instructions
CPU L1 data cache	16 kB
CPU L2 cache	2 MB (on-Die, ECC, ATC, full-speed)
System memory	1024 MB
System memory speed	533 MHz
Disk drive	WD, WD800JD-75JNC0 (74 GB, IDE)

Appendix D Platform specifications

D.4 Dell Precision 380 (Dual core desktop)

Component	Description
CPU type	Intel Core 2 Duo E6700 (Conroe)
CPU original clock speed	2667 MHz
CPU L1 code cache	32 kB per core
CPU L1 data cache	32 kB per core
CPU L2 cache	4 MB (on-Die, ASC, full-speed)
System memory	2048 MB
System memory speed	667 MHz
Disk drive	2x WD Raptor (74 GB, S-ATA, RAID0)

D.5 Custom made PC (Single core desktop)

Component	Description
CPU type	AMD Athlon 64 3700+ (Core San Diego)
CPU original clock speed	2210 MHz
CPU L2 cache	2 MB
System memory	1024 MB
Disk drive	2x WD Raptor (36 GB, IDE, RAID0)

Bibliography

[1] Jean-Marie Zogg. *GPS Essentials of Satellite Navigation*. u-blox AG, 2009. ISBN 978-3033021396.

[2] Navstar. Navstar GPS Space Segment/User Segment L1C Interfaces (Draft IS-GPS-800), September 2008. URL http://www.losangeles.af.mil.

[3] Elliot D. Kaplan and Christopher J. Hegarty. *Understanding GPS – Principles and applications*. Artech House, Inc., 2nd edition, 2006. ISBN 978-0890067932.

[4] Pratap Misra and Per Enge. *Global Positioning System: Signals, Measurements, and Performance*. Ganga-Jamuna Press, 2nd edition, 2006. ISBN 0-9709544-1-7.

[5] InsideGNSS. Russia to put 8 CDMA signals on 4 GLONASS frequencies, March 2010. URL http://www.insidegnss.com/node/1997.

[6] Federal Space Agency Information-Analytical Centre. Internet, 2010. URL http://www.glonass-ianc.rsa.ru.

[7] European Commission. Galileo System. Internet, May 2010. URL http://ec.europa.eu/enterprise/policies/space/galileo/.

[8] Kai Borre and Dennis M. Akos. *A Software-Defined GPS and Galileo Receiver: A Single-Frequency Approach*. Birkhäuser Boston, 1st edition, 2006. ISBN 978-0817643904.

[9] Stephan Esterhuizen. *The Design, Construction, and Testing of a Modular GPS Bistatic Radar Software Receiver for Small Platforms*. PhD thesis, University of Colorado, 2006.

[10] Bradford W. Parkinson and James J. Spilker Jr., editors. *Global Positioning System: Theory and Applications, Vol. I, II*, volume 163–164 of *Progress in Astronautics and Aeronautics*. American Institute of Aeronautics, Inc., Washington DC, 1996.

[11] James B. Y. Tsui. *Fundamentals of Global Positioning System Receivers - A Software Approach*. John Wiley and Sons, Inc., 2nd edition, 2005. ISBN 978-0471706472.

[12] Navstar. Navstar GPS Space Segment/Navigation User Interfaces (IS-GPS-200 Rev. D), March 2006. URL http://www.losangeles.af.mil.

[13] Thomas Pany. *Navigation Signal Processing for GNSS Software Receivers.* Artech House, Inc., 1st edition, 2010. ISBN 978-1608070275.

[14] Thomas Pany and Bernd Eissfeller. Code and Phase Tracking of Generic PRN Signals with Sub-Nyquist Sample Rates. *Navigation, Journal of The Institute of Navigation*, 51(2):143–159, 2004.

[15] John Raquet. GNSS receiver design. Lecture notes, Department of Geomatics Engineering, The University of Calgary, 2006.

[16] D. J. R. van Nee and A. J. R. M. Coenen. New fast GPS code acquisition technique using FFT. *Electronics Letters*, 27(2):158–160, 1991.

[17] David Akopian. Fast FFT based GPS satellite acquisition methods. *IEEE Proc. Radar Sonar Navigation*, 152(4):277–286, 2005.

[18] A. J. van Dierendonck. *Global Positioning System: Theory and Applications Volume I*, chapter GPS Receivers, pages 329–407. American Institute of Aeronautics, Inc., 1996.

[19] Roger L. Peterson, Roger E. Ziemer, and David E. Borth. *Introduction to Spread Spectrum Communications.* Prentice Hall, April 1995. ISBN 978-0024316233.

[20] R. I. Lackey and D. W. Upmal. Speakeasy: The military software radio. *IEEE Communications Magazine*, 33(5):56–61, 1995. doi: 10.1109/35.392998.

[21] A. J. R. M. Coenen and D. J. R. van Nee. Novel fast GPS/Glonass code acquisition technique using low update rate FFT. *Electronics Letters*, 28(9):863–865, April 1992.

[22] Dennis M. Akos. *A Software Radio Approach to Global Navigation Satellite System Receiver Design.* PhD thesis, Ohio University, August 1997.

[23] Dennis M. Akos, Per-Ludvig Normark, Per Enge, Andreas Hansson, and Andreas Rosenlind. Real-Time GPS software radio receiver. In *ION NTM*, Long Beach, CA, January 2001.

[24] Philip G. Mattos. A low-cost hand-held GPS navigation system receiver. In *Fourth International Conference on Satellite Systems for Mobile Communications and Navigation*, pages 217–221, 1989.

[25] GPS World. Internet, 2010. URL http://www.gpsworld.com.

[26] Altera. Nios II Processor. Internet, 2010. URL http://www.altera.com/products/ip/processors/nios2.

[27] J. Mitola. What is a Software Radio? Internet, 2005. URL http://web.archive.org/web/20050315234159/http://ourworld.compuserve.com/homepages/jmitola/whatisas.htm.

[28] *IA-32 Intel Architecture Software Developer's Manual Volume 3B: System Programming Guide*. Intel Corporation, March 2006. URL http://www.intel.com/products/processor/manuals/.

[29] Gregory W. Heckler and James L. Garrison. SIMD correlator library for GNSS software receiver. *GPS Solutions*, 10:269–276, 2006.

[30] Brent M. Ledvina, Steven P. Powell, and Paul M. Kintner. A 12-Channel Real-Time GPS L1 Software Receiver. In *ION NTM*, pages 767–782, Anaheim, CA, January 2003.

[31] Andrew J. Viterbi. *CDMA, Principles of Spread Spectrum Communication*. Addison-Wesley, April 1995. ISBN 978-0201633740.

[32] Brent M. Ledvina, Mark L. Psiaki, Steven P. Powell, and Paul M. Kintner. Real-time software receiver. U.S. Patent US0227856, October 2006.

[33] Thomas Pany, Sung Wook Moon, Markus Irsigler, Bernd Eissfeller, and Karl Fürlinger. Performance assessment of an under sampling SWC receiver for simulated high bandwidth GPS/Galileo signals and real signals. In *ION GPS/GNSS*, Portland, OR, September 2003.

[34] Shahin Charkhandeh, Mark Petovello, R. Watson, and Gérard Lachapelle. Implementation and Testing of a Real-Time Software-Based GPS Receiver for x86 Processors. In *ION NTM*, Monterey, CA, January 2006.

[35] Alexander Fridmann and Serguei Semenov. Architectures of software GPS receivers. *GPS Solutions*, 3(4):58–64, 2000.

[36] Per-Ludvig Normark and Christian Stahlberg. Spread spectrum signal processing. U.S. Patent WO2004/036238, October 2003.

[37] Mark Petovello and Gérard Lachapelle. An Efficient New Method for Doppler Removal and Correlation with Application to Software-Based GNSS Receivers. In *ION GNSS*, Fort Worth, TX, September 2006.

[38] Gregory W. Heckler and James L. Garrison. Architecture of a Reconfigurable Software Receiver. In *ION GNSS*, Long Beach, CA, September 2004.

[39] Yu-Hsuan Chen and Jyh-Ching Juang. A GNSS software receiver approach for the processing of intermittent data. In *ION GNSS*, Fort Worth, TX, 2007.

[40] Mark L. Psiaki. Real-Time generation of Bit-Wise parallel representation of Over-Sampled PRN code. *IEEE Trans. on Wireless Communication*, 5(3), March 2006.

[41] Jin Tian, Qin HongLei, Zhu JunJie, and Liu Yang. Real-time GPS Software Receiver Correlator Design. In *ChinaCom2007*, Shanghai, China, August 2007.

[42] Brent M. Ledvina, Mark L. Psiaki, D. J. Sheinfeld, Alessandro P. Cerruti, Steven P. Powell, and Paul M. Kintner. A Real-Time GPS civilian L1/L2 software receiver. In *ION GNSS*, Long Beach, CA, September 2004.

[43] Chris Hegarty Chun Yang and Micahel Tran. Acquisition of the GPS L5 Signal Using Coherent Combining of I5 and Q5. In *ION GNSS*, Long Beach, CA, September 2004.

[44] J.W. Cooley and J.W. Tukey. An algorithm for the machine computation of the complex fourier series. *Math. Computations*, 19:297–301, April 1965.

[45] Dennis M. Akos, Per-Ludvig Normark, Andreas Hansson, Andreas Rosenlind, Christian Stahlberg, and Fredrik Svensson. Global Positioning System Software Receiver (gpSrx) Implementation in Low Cost/Power Programmable Processors. In *ION GPS*, Salt Lake City, UT, September 2001.

[46] Mark L. Psiaki, Dennis M. Akos, and Jonas Thor. A Comparison of Direct Radio Frequency Sampling and Conventional GNSS Receiver Architectures. *Navigation: Journal of The Institute of Navigation*, 52(2), 2005.

[47] Nicolaj Bertelen, Kai Borre, and Peter Rinder. The GPS Code Software Receiver at Aalborg University. In *2nd ESA Workshop on Satellite navigation*, pages 373–380, 2004.

[48] Aleksandar Jovancevic, Andrew Brown, Suman Ganguly, Michael Kirchner, and Slavisa Zigic. Reconfigurable Dual Frequency Software GPS Receiver and Applications. In *ION GPS*, Salt Lake City, UT, September 2001.

[49] C. Ma, Gérard Lachapelle, and M. Elizabeth Cannon. Implementation of a software GPS receiver. In *ION GNSS*, Long Beach, CA, September 2004.

[50] John J. Schamus, James B. Y. Tsui, and David M. Lin. Real-Time Software GPS Receiver. In *ION GPS*, Portland, OR, September 2002.

[51] Thomas Pany, Günther Hein, Sung Wook Moon, and Daniel Sanroma. ipexSR: A PC Based Software GNSS Receiver Completely Developed in Europe. In *ENC GNSS*, Copenhagen, DK, 2002.

[52] Paulo S. R. Diniz, Eduardo A. B da Silva, and Sergio L. Netto. *Digital Signal Processing, System Analysis and Design*. Cambridge University Press, 1st edition, 2002. ISBN 978-0521025133.

[53] *Intel Integrated Performance Primitives v5.0 for Windows on Intel Pentium Processors*. Intel Corporation, 2007. URL http://www.intel.com.

[54] Matteo Frigo and Steven G. Johnson. FFTW: Fastest Fourier Transform in the West. Internet, 2007. URL http://www.fftw.org.

[55] *Pentium Processors with MMX Technology.* Intel Corporation, 2010. URL http://www.intel.com.

[56] Grégoire Waelchli, Marcel Baracchi-Frei, Cyril Botteron, and Pierre-André Farine. Performances of a new correlation algorithm for a platform-independent GPS software receiver. In *ION ITM*, Anaheim CA, 2009.

[57] Hubert Nguyen. *CPU Gems 3.* Addison-Wesley Professional, August 2007. ISBN 978-0321515261.

[58] John D. Owenns, Mike Housten, David Luebke, Simon Green, John E. Stone, and James C. Philips. GPU Computing. *Proceedings of IEEE*, 96(5):879–899, May 2008.

[59] Chris Harris, Karen Haines, and Lister Staveley-Smith. GPU Accelerated Radio Astronomy Signal Convolution. *Experimental Astronomy*, 22(1–2):129–141, October 2008. doi: 10.1007/s10686-008-9114-9.

[60] Thomas Hobiger, Tadahiro Gotoh, Jun Amagai, Yasuhiro Koyama, and Tetsuro Kondo. A GPU based real-time GPS software receiver. *GPS Solutions*, August 2009. doi: 10.1007/s10291-009-135-2.

[61] NVIDIA. Computer Unified Device Architecture CUDA. Internet, 2010. URL http://developer.nvidia.com/.

[62] Clemens Bürgi, Marcel Baracchi, and Grégoire Waelchli. A method of processing a digital signal derived from a direct-sequence spread spectrum signal and a receiver for carrying out the method. European Patent No. 09405207.3 - 2411, November 2009.

[63] Clemens Bürgi, Marcel Baracchi, and Grégoire Waelchli. A method of processing a digital signal derived from a direct-sequence spread spectrum signal and a receiver. U.S. Patent, Application No. 12/694,145, January 2010.

[64] *EZ-USB FX2LP USB Microcontroller Datasheet.* Cypress Semiconductor Corporation, January 2006. URL http://www.cypress.com. Document number: 38-08032, Rev. K.

[65] *Windows Administation - Inside the Windows Vista Kernel: Part 1.* Microsoft Corporation, Mark Russinovich, 2007. URL http://technet.microsoft.com/en-us/magazine/2007.02.vistakernel.aspx.

[66] *Coaxial Power Splitter ZN2PD2-50+.* Mini-Circuits, 2010. URL http://www.minicircuits.com/pdfs/ZN2PD2-50+.pdf.

[67] *Process Explorer.* Microsoft Corporation, Mark Russinovich, 2009. URL http://technet.microsoft.com/en-us/sysinternals/bb896653.aspx.

[68] Robert C. Dixon. *Spread Spectrum Systems*. John Wiley & Sons, 2nd edition, 1984. ISBN 0-471-88309-3.

[69] Floyd M. Gardner. *Phaselock Techniques*. Wiley-Interscience, 2nd edition, April 1979. ISBN 978-0471042945.

Die VDM Verlagsservicegesellschaft sucht für wissenschaftliche Verlage abgeschlossene und herausragende

Dissertationen, Habilitationen, Diplomarbeiten, Master Theses, Magisterarbeiten usw.

für die kostenlose Publikation als Fachbuch.

Sie verfügen über eine Arbeit, die hohen inhaltlichen und formalen Ansprüchen genügt, und haben Interesse an einer honorarvergüteten Publikation?

Dann senden Sie bitte erste Informationen über sich und Ihre Arbeit per Email an *info@vdm-vsg.de*.

Sie erhalten kurzfristig unser Feedback!

VDM Verlagsservicegesellschaft mbH
Dudweiler Landstr. 99
D - 66123 Saarbrücken
www.vdm-vsg.de

Telefon +49 681 3720 174
Fax   +49 681 3720 1749

Die VDM Verlagsservicegesellschaft mbH vertritt

Printed by Books on Demand GmbH, Norderstedt / Germany